Ms.45

Sequential Data in Biological Experiments

SEQUENTIAL DATA IN BIOLOGICAL EXPERIMENTS

An introduction for research workers

E.A. Roberts

Sydney Water Board

CHAPMAN & HALL

London · New York · Tokyo · Melbourne · Madras

Published by Chapman & Hall, 2–6 Boundary Row, London SE1 8HN

Chapman & Hall, 2–6 Boundary Row, London SE1 8HN, UK

Chapman & Hall, 29 West 35th Street, New York NY10001, USA

Chapman & Hall Japan, Thomson Publishing Japan, Hirakawacho Nemoto Building, 7F, 1-7-11 Hirakawa-cho, Chiyoda-ku, Tokyo 102, Japan

Chapman & Hall Australia, Thomas Nelson Australia, 102 Dodds Street, South Melbourne, Victoria 3205, Australia

Chapman & Hall India, R. Seshadri, 32 Second Main Road, CIT East, Madras 600 035, India

First edition 1992

© 1992 E.A. Roberts

Typeset in 10/12 Times by Interprint, Malta
Printed in Great Britain by T.J. Press Ltd, Padstow, Cornwall

ISBN 0 412 414104

A catalogue record for this book is available from the British Library

Library of Congress Cataloging-in-Publication data available

To Judith

Contents

Preface

There are many excellent books on general statistical methods in agricultural and biological research. These books cover a broad range of methods without going into detail on specialized topics. A number of topics including regression analysis, design of experiments, biological assay and categorical analysis have received in-depth treatment in specialized texts. Little appears in standard textbooks on experiments in which observations form sequences.

The live weights of animals during a long-term experiment provide a familiar example of data forming a sequence, but many others occur: for example, moisture content of segments of soil cores, successive counts of insects in an orchard and hormone levels in blood over a period. Correlations are likely to be found among the observations in all these examples. The book by Goldstein (1979) provided the first systematic coverage of the principles involved in longitudinal studies, but is mainly concerned with observational studies on humans.

The main aims of this book are to provide research workers with methods of analysing data from comparative experiments with sequential observations and to demonstrate special features of the design of such experiments. These aims are achieved by working through sets of data.

All the sets of data, except those used in Chapters 1 and 10, came from experiments carried out by research workers of the New South Wales Department of Agriculture. The results of these experiments have been given in agricultural journals and references are given for these when they occur in the book. The data in Chapter 1 were obtained from Mr A.L.C. Wallace of CSIRO by Mr A. Gleeson, who was interested in the subject matter; the set in Chapter 10 was brought to my attention by Dr J.C. Evans, who was interested in the method of analysis.

An appropriate analysis for sequential data was given by Wishart (1938) for growth situations and was advocated by Rowell and Walters (1976) in general. The method is based on analysing contrasts in the observations on an individual. Although the method has been used in agricultural experiments, e.g. Yates *et al.* (1964), it has not been widely accepted by research workers because the results of the analyses are difficult for them to interpret in terms of the measured response. Researchers commonly prefer to analyse the data separately each time although this does not provide any test of how treatment effects vary over time (Rowell and Walters, 1976). However, analysis of contrasts over time does, in fact, provide appropriate tests for change in treatment effects in time (Evans and Roberts, 1979). Hence, the approaches of the research worker and the statistician are complementary. They are combined in the method used in this book: analysis of the data at the individual time-points provides the estimates of the treatment effects over the period of the experiment; analysis of contrasts over time provides the tests of trends in these estimates.

Many people have contributed in many ways to the development of this book. I express my gratitude to them all. The basic method used to analyse data from comparative experiments is given in Chapters 2 and 3. The method was developed with Dr G.M. Lodge, then Research Agronomist, to analyse the data in Lodge and Roberts (1979), part of which is used in Chapter 5. Dr J.C. Evans and Mr G.H. O'Neill helped considerably with the statistical methods. All the analyses were performed with the computer program REG written by Dr A.R. Gilmour. Mrs Joy Raison, Mrs Jean McGarry, Miss Margaret de Koning and Miss Vicki Harris assisted with the analyses. Mrs Raison also wrote the program to calculate orthogonal polynomials through the origin. I am particularly grateful to my former colleague Mr Cameron Kirton who kindly read the whole manuscript and found a number of omissions and obscurities.

1 Introduction

1.1 INTRODUCTION

Most of this book is concerned with the analysis of experiments in which a number of observations are taken on each unit in the experiment. The special features of the data obtained in such experiments and of the analyses needed, will be considered in this chapter after brief reviews of aspects of matrix algebra and of the statistical methods used in the book. The notation of matrices is particularly convenient for the algebraic expressions which are frequently needed to define the values predicted by regression equations and their variances.

1.2 MATRICES

Any rectangular array of numbers is called a matrix. If the array has r rows and s columns it is called an r by s ($r \times s$) matrix. Bold capital letters are usually used to denote matrices and double subscripts are used to locate an element in the matrix. Thus, for example,

$$\underset{2 \times 3}{A} = \begin{bmatrix} a_{11} & a_{12} & a_{13} \\ a_{21} & a_{22} & a_{23} \end{bmatrix} \quad \text{and} \quad \underset{3 \times 2}{B} = \begin{bmatrix} b_{11} & b_{12} \\ b_{21} & b_{22} \\ b_{31} & b_{32} \end{bmatrix}.$$

1.2.1 Transpose of a matrix

The transpose of the matrix A is written A^{T} and is obtained by writing the first row of A as the first column of A^{T} and so on. Therefore, if

$$\underset{2 \times 3}{A} = \begin{bmatrix} a_{11} & a_{12} & a_{13} \\ a_{21} & a_{22} & a_{23} \end{bmatrix} \quad \text{then} \quad \underset{3 \times 2}{A^{\mathrm{T}}} = \begin{bmatrix} a_{11} & a_{21} \\ a_{12} & a_{22} \\ a_{13} & a_{23} \end{bmatrix}.$$

1.2.2 Addition of matrices

Two matrices, A and B, can only be added if they have the same dimensions. Their sum is obtained by adding corresponding elements. For example, if

$$\underset{2 \times 3}{A} = \begin{bmatrix} 2 & 1 & 4 \\ -1 & 6 & 2 \end{bmatrix} \quad \text{and} \quad \underset{2 \times 3}{B} = \begin{bmatrix} 0 & -1 & 1 \\ 6 & 0 & -3 \end{bmatrix}$$

then

$$A + B \atop 2 \times 3 = \begin{bmatrix} 2+0 & 1-1 & 4+1 \\ -1+6 & 6+0 & 2-3 \end{bmatrix} = \begin{bmatrix} 2 & 0 & 5 \\ 5 & 6 & -1 \end{bmatrix}.$$

1.2.3 Multiplication of a matrix by a real number

The product of a real number c and a matrix A is the matrix in which every element of A has been multiplied by c.

1.2.4 Product of matrices

The rules for matrix multiplication are not as obvious as those for addition, and multiplication requires practice. The product of A $(r \times m)$ and B $(n \times s)$ is written AB and has dimensions $r \times s$. It is obtained by series of row–column multiplications which are best described by examples.

Example 1

Let A and B be

$$A = \begin{bmatrix} 2 & 0 \\ 1 & 4 \end{bmatrix}, \quad B = \begin{bmatrix} 5 & 2 \\ -1 & 3 \end{bmatrix}$$

then

$$AB = \begin{bmatrix} 2(5)+0(-1) & 2(2)+0(3) \\ 1(5)+4(-1) & 1(2)+4(3) \end{bmatrix} = \begin{bmatrix} 10 & 4 \\ 1 & 14 \end{bmatrix}.$$

The ijth element of AB is the sum of the cross products of the elements of row i of A and the elements of column j of B.

Example 2

Let A and B be

$$A = \begin{bmatrix} 2 & 1 \\ 1 & -1 \\ 3 & 2 \end{bmatrix}, \quad B = \begin{bmatrix} 4 & -1 & 1 \\ 2 & 0 & 2 \end{bmatrix}$$

then

$$AB = \begin{bmatrix} 2(4)+1(2) & 2(-1)+1(0) & 2(1)+1(2) \\ 1(4)-1(2) & 1(-1)+1(0) & 1(1)-1(2) \\ 3(4)+2(2) & 3(-1)+2(0) & 3(1)+2(2) \end{bmatrix} = \begin{bmatrix} 10 & -2 & 4 \\ 2 & -1 & -1 \\ 16 & -3 & 7 \end{bmatrix}$$

and

$$BA = \begin{bmatrix} 4 & -1 & 1 \\ 2 & 0 & 2 \end{bmatrix} \begin{bmatrix} 2 & 1 \\ 1 & -1 \\ 3 & 2 \end{bmatrix}$$

$$= \begin{bmatrix} 4(2)-1(1)+1(3) & 4(1)-1(-1)+1(2) \\ 2(2)+0(1)+2(3) & 2(1)+0(-1)+2(2) \end{bmatrix}$$

$$= \begin{bmatrix} 10 & 7 \\ 10 & 6 \end{bmatrix}.$$

Example 2 shows that, in general, AB does not equal BA. Two matrices A and B can only be multiplied when the rows of A contain the same number of elements as the columns of B, i.e. the number of columns in A equals the number of rows in B.

1.2.5 The identity matrix, I, and the inverse of a matrix

The identity matrix, I, is a square matrix in which the elements on the main diagonal all equal one and all other elements are zero, i.e.

$$\underset{7\times 7}{I} = \begin{bmatrix} 1 & 0 & 0 & 0 & 0 & 0 & 0 \\ 0 & 1 & 0 & 0 & 0 & 0 & 0 \\ 0 & 0 & 1 & 0 & 0 & 0 & 0 \\ 0 & 0 & 0 & 1 & 0 & 0 & 0 \\ 0 & 0 & 0 & 0 & 1 & 0 & 0 \\ 0 & 0 & 0 & 0 & 0 & 1 & 0 \\ 0 & 0 & 0 & 0 & 0 & 0 & 1 \end{bmatrix}.$$

The inverse of a square matrix A is written A^{-1} and is defined in terms of the identity matrix as follows. If a matrix A^{-1} can be found such that

$$A^{-1}A = AA^{-1} = I,$$

A^{-1} is called the inverse of A. A^{-1} is defined in this conditional manner because many square matrices do not have inverses. Matrices that do not have inverses are said to be singular. Singularity indicates that one or more columns of the matrix can be expressed as a linear function of the other columns, i.e. the columns are not independent.

1.2.6 Determinant of a matrix

The determinant of a matrix A is written $|A|$. There is no simple definition of $|A|$ for an $n \times n$ matrix; however, we only need note that the value of $|A|$ depends on the structure of A as well as on the size of its elements. In particular, $|A| = 0$ when any column of A is a linear function of the other columns, i.e. when A is singular.

1.3 SIMPLE LINEAR REGRESSION

The regression equation relating observations y_i to the chosen values, x_i, of another variable can be written in the form

$$\hat{y}_i = b_0 + b_1 x_i \qquad i = 1, \ldots, n \qquad (1.1)$$

where $\hat{y}_i$ is the expected or fitted value of y, the dependent variable, corresponding to the value x_i of the independent variable; b_0 is the constant term; and b_1 is the regression coefficient relating the values of y and x.

The estimates of b_0 and b_1 are obtained by the method of least squares; this will be referred to as regressing the values of y on the values of x. In matrix notation these estimates are:

$$\hat{\beta} = \begin{bmatrix} b_0 \\ b_1 \end{bmatrix} = (X^T X)^{-1} X^T Y$$

where,

$$Y = \begin{bmatrix} y_1 \\ y_2 \\ y_3 \\ . \\ . \\ . \\ y_n \end{bmatrix} \quad \text{and} \quad X = \begin{bmatrix} 1 & x_1 \\ 1 & x_2 \\ 1 & x_3 \\ . & \\ . & \\ . & \\ 1 & x_n \end{bmatrix}$$

The values x_i will often be coded values of observed or chosen values of a variable. For example, if observations were taken on four occasions at fortnightly intervals, the fortnights would be considered to form a variable, t, with levels 1, 2, 3 and 4. Coded values of t would be -3, -1, 1, 3 which are the values of $t - \bar{t}$, multiplied by 2. The regression Equation (1.1) will usually be written without subscripts.

1.4 MULTIPLE LINEAR REGRESSION

Similarly, the regression equation relating the values of a dependent variable, y, and independent variables $x_1, \ldots, x_n$ will be written

$$\hat{y} = b_0 + b_1 x_1 + \cdots + b_n x_n, \tag{1.2}$$

where b_0 is the constant term and $b_1, \ldots, b_n$ are partial regression coefficients relating y and $x_1, \ldots, x_n$. The estimates of the coefficients $b_0, \ldots, b_n$ are obtained by regressing y on $x_1, \ldots, x_n$ in multiple regression. It is important to remember that, in general, the value obtained for any coefficient in multiple regression and the sum of squares associated with it depend on which other variables are included in the regression. Hence, if an initial analysis indicates that one of the x-variables is not important and this variable is omitted, the regression must be recalculated using the remaining variables to obtain the required estimates; this is a consequence of relations between the values of the x-variables. The values of the x-variables can be chosen or transformed to avoid this problem; the x-variables are then said to be orthogonal to one another.

1.5 POLYNOMIAL REGRESSION

Polynomial equations are often used to approximate or smooth the relation between a response, y, and a quantitative variable, x, such as time or concentration. The linear model for a polynomial of degree n is

$$\hat{y} = b_0 + b_1 x + b_2 x^2 + \cdots + b_n x^n.$$

This is the same form as Equation (1.2) with $x_1 = x$, $x_2 = x^2$, $x_3 = x^3$, etc., and the coefficients $b_0, \ldots, b_n$ are estimated by regressing the values of y on the values of x, x^2, x^3, etc., in multiple regression. The values in columns 1 and 2 of Table 1.1 will be used as a simple example of polynomial regression. The y-values in column 1 are 1000 times the differences in live weight in natural logarithms between sheep grazing at two stocking rates. (They have been extracted from the experiment discussed in Chapter 3.) The sheep were weighed about every 40 days.

The values of y are plotted against weighing time in Figure 1.1. It appears that a quadratic curve would be required to approximate the trend in the y-values with time. A quadratic is a polynomial of degree 2. A polynomial of degree 4 (quartic) was fitted in case the higher-order terms were needed, and to illustrate the principles.

The regression fitted was

$$\hat{y} = b_0 + b_1(t-7) + b_2(t-7)^2 + b_3(t-7)^3 + b_4(t-7)^4$$

or

$$y = b_0 + b_1 t + b_2 t^2 + b_3 t^3 + b_4 t^4,$$

Table 1.1   *y*-values at 13 weighings, *t*, and variables derived from *t* for polynomial regression of *y* on *t*

y	*t*	Polynomial values of $(t-7)$ up to $(t-7)^4$				Orthogonal polynomial values for *t* of order 1 to 4			
		$(t-7)$	$(t-7)^2$	$(t-7)^3$	$(t-7)^4$	ξ_1	ξ_2	ξ_3	ξ_4
32.8	1	−6	36	−216	1296	−6	22	−11	99
42.0	2	−5	25	−125	625	−5	11	0	−66
81.4	3	−4	16	−64	256	−4	2	6	−96
69.0	4	−3	9	−27	81	−3	−5	8	−54
94.3	5	−2	4	−8	16	−2	−10	7	11
81.3	6	−1	1	−1	1	−1	−13	4	64
102.5	7	0	0	0	0	0	−14	0	84
67.3	8	1	1	1	1	1	−13	−4	64
74.8	9	2	4	8	16	2	−10	−7	11
62.2	10	3	9	27	81	3	−5	−8	−54
91.5	11	4	16	64	256	4	2	−6	−96
94.4	12	5	25	125	625	5	11	0	−66
70.8	13	6	36	216	1296	6	22	11	99

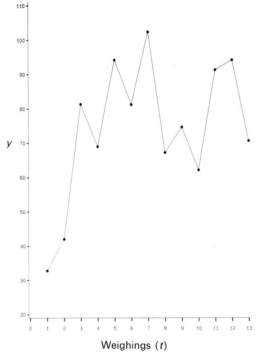

Figure 1.1   *y*-values of Table 1.1 plotted against time.

writing t for $(t-7)$, i.e. $t-\bar{t}$. The values of these variables are given in Table 1.1 together with others needed later. The results of the regression analysis are summarized neatly in the format of an analysis of variance as shown in Table 1.2(a).

Table 1.2 Analyses of variance for polynomial regression of the y-values of Table 1.1 on t

(a) Regression on t, t^2, t^3, t^4 coding t as $t-7$

Source of variation	DF	Sums of squares	Mean squares	F	P
Regression on:					
t	1	1147.52	1147.52	4.93	0.057
t^2 after t	1	1260.25	1260.25	5.42	0.048
t^3 after t and t^2	1	638.42	638.42	2.74	0.136
t^4 after t, t^2 and t^3	1	88.09	88.0	<1	
Residual	8	1861.56	232.70		
Total	12	4995.84			

The values of the coefficients estimated in the regression were:

Variable	Coefficient	Standard error
Constant	82.2628	8.05
t	-1.8910	2.89
t^2	-0.0529	1.25
t^3	-0.1761	0.106
t^4	-0.0210	0.034

(b) Regression on t and t^2

Source of variation	DF	Sums of squares	Mean squares	F	P
t	1	1147.52	1147.52	4.43	0.061
t^2 after t	1	1260.25	1260.25	4.87	0.052
Residual	10	2588.10	258.81		

Variable	Coefficient	Standard error
Constant	85.2846	6.72
t	2.5110	1.19
t^2	0.7934	0.360

The y-values were next regressed on t and t^2 because the additional amounts of variation accounted for by including t^4 in the regression after t, t^2 and t^3 and by including t^3 after t and t^2 were not significant. The results of this regression analysis are summarized in Table 1.2(b). A notable feature of these analyses is the change in the values of the regression coefficients, particularly in b_1 which changed from -1.8910 in the presence of t^3 and t^4 to 2.5110 in their absence.

The estimated variances and covariances of the regression coefficients are given by

$$\text{var}(\hat{\boldsymbol{\beta}}) = s^2 (\boldsymbol{X}^T \boldsymbol{X})^{-1}$$

where $s^2 = 258.81$, the residual mean square in the analysis of variance. The calculations are set out below:

$$
\boldsymbol{X} = \begin{bmatrix}
1 & -6 & 36 \\
1 & -5 & 25 \\
1 & -4 & 16 \\
1 & -3 & 9 \\
1 & -2 & 4 \\
1 & -1 & 1 \\
1 & 0 & 0 \\
1 & 1 & 1 \\
1 & 2 & 4 \\
1 & 3 & 9 \\
1 & 4 & 16 \\
1 & 5 & 25 \\
1 & 6 & 36
\end{bmatrix},
\quad \text{therefore} \quad
\boldsymbol{X}^T \boldsymbol{X} = \begin{bmatrix}
13 & 0 & 182 \\
0 & 182 & 0 \\
182 & 0 & 4550
\end{bmatrix}
$$

and

$$
(\boldsymbol{X}^T \boldsymbol{X})^{-1} = \begin{bmatrix}
25/143 & 0 & -1/143 \\
0 & 1/182 & 0 \\
-1/143 & 0 & 1/2002
\end{bmatrix}.
$$

So that

$$
\text{var}(\hat{\boldsymbol{\beta}}) = 258.81 \begin{bmatrix}
25/143 & 0 & -1/143 \\
0 & 1/182 & 0 \\
-1/143 & 0 & 1/2002
\end{bmatrix}
$$

$$= \begin{bmatrix} 45.2465 & 0 & -1.8099 \\ 0 & 1.4220 & 0 \\ -1.8099 & 0 & 0.1293 \end{bmatrix}.$$

The diagonal elements are the variances of the estimates and the off-diagonal elements are their covariances. Hence

$$\text{standard error of } b_0 = 45.2465^{1/2} = 6.73$$
$$\text{standard error of } b_1 = 1.4220^{1/2} = 1.19$$
$$\text{and standard error of } b_2 = 0.1293^{1/2} = 0.360.$$

An estimate such as a mean or a regression coefficient is usually rounded to a convenient interval near one-tenth of its standard error to avoid implying greater precision in the estimate than is warranted by the data. Therefore, the fitted equation would be reported as

$$\hat{y} = \underset{(\pm 6.73)}{85.3} + \underset{(\pm 1.19)}{2.5t} - \underset{(\pm 0.36)}{0.79t^2}$$

although the fitted values are usually calculated before the coefficients are rounded.

Fitted values are obtained by substituting the values of t and t^2 in the fitted equation or equivalently as the matrix product $X\hat{\beta}$ as follows:

$$X\hat{\beta} = \begin{bmatrix} 1 & -6 & 36 \\ 1 & -5 & 25 \\ 1 & -4 & 16 \\ 1 & -3 & 9 \\ 1 & -2 & 4 \\ 1 & -1 & 1 \\ 1 & 0 & 0 \\ 1 & 1 & 1 \\ 1 & 2 & 4 \\ 1 & 3 & 9 \\ 1 & 4 & 16 \\ 1 & 5 & 25 \\ 1 & 6 & 36 \end{bmatrix} \begin{bmatrix} 85.2846 \\ 2.5110 \\ 0.7934 \end{bmatrix} = \begin{bmatrix} 41.6560 \\ 52.8945 \\ 62.5462 \\ 70.6110 \\ 77.0890 \\ 81.9802 \\ 85.2846 \\ 87.0022 \\ 87.1330 \\ 85.6769 \\ 82.6341 \\ 78.0044 \\ 71.7879 \end{bmatrix}.$$

These fitted values were used to plot the curve in Figure 1.2. Finally, writing Y for the column of fitted values, the variances and covariances of

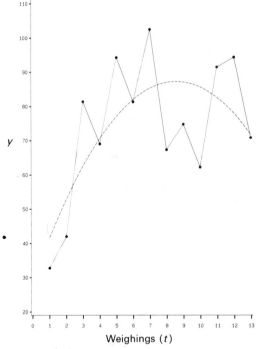

Figure 1.2   Quadratic regression fitted to the y-values given in Table 1.1.

the fitted values are given by

$$\text{var}(Y) = X\,\text{var}(\hat{\beta})X^{\text{T}}$$

These values can be used to obtain confidence limits for the fitted values and confidence bands for the fitted polynomial.

1.5.1 Orthogonal polynomials

The values of the orthogonal polynomials of order 1 to 4 for 13 equally spaced intervals are also given in Table 1.1. The polynomial of order 0 is a column of ones. Two columns of a matrix are orthogonal when the sum of the products of corresponding elements is equal to zero.

In this example the values of the orthogonal polynomials are:

$$\xi_0 = 1$$
$$\xi_1 = (t - \bar{t})$$
$$\xi_2 = (t - \bar{t})^2 - 14$$
$$\xi_3 = (t - \bar{t})^3 - 25(t - \bar{t})$$
$$\xi_4 = 7(t - \bar{t})^4/12 - 247(t - \bar{t})^2/7 + 84.$$

In general for equally spaced intervals,

$$\xi_{r+1} = \xi_1 \xi_r - r^2(n^2 - r^2)\xi_{r-1}/(16r^2 - 4). \tag{1.3}$$

When these values are used in place of the values of the powers of $(t - \bar{t})$ in polynomial regression, the values of the coefficients obtained do not depend on which variables are included in the regression. Also, importantly for the type of experiment considered in this book, the constant term in the regression equation is equal to the mean of the y-values. The analysis of variance for regression of the y-values of Table 1.1 on the values of ξ_1, ξ_2, ξ_3 and ξ_4 given in Table 1.3(a) indicated that ξ_3 and ξ_4 were non-significant and could be omitted from the regression. The results of the regression on ξ_1 and ξ_2 are given in Table 1.3(b). However, the y-values did not need to be regressed on the values of ξ_1 and ξ_2 alone to obtain the required

Table 1.3 Analyses of variance for polynomial regression using orthogonal polynomials

(a) Regression on ξ_1, ξ_2, ξ_3 and ξ_4 of Table 1.1

Source	DF	SS	MS	F	P
ξ_1	1	1147.52	1147.52	4.93	0.057
ξ_2	1	1260.25	1260.25	5.42	0.048
ξ_3	1	638.42	638.42	2.74	0.136
ξ_4	1	88.09	88.09	0.38	0.555
Residual	8	1861.56			

Variable	Coefficient	Standard error
Constant	74.1769	4.23
ξ_1	2.5110	1.13
ξ_2	−0.7934	0.341
ξ_3	1.0565	0.638
ξ_4	−0.0360	0.0585

(b) Regression on ξ_1 and ξ_2

Source	DF	SS	MS	F	P
ξ_1	1	1147.52	1147.52	4.43	0.061
ξ_2	1	1260.25	1260.25	4.87	0.052
Residual	10	2588.07	258.81		

Variable	Coefficient	Standard error
Constant	74.1769	4.46
ξ_1	2.5110	1.19
ξ_2	−0.7934	0.360

regression because the values of b_1 and b_2 in that regression are the same as those estimated in the regression on ξ_1, ξ_2, ξ_3 and ξ_4. Standard errors of b_1 and b_2 are based on the residual which will be estimated on ten degrees of freedom when ξ_3 and ξ_4 are omitted. The sum of squares for this residual is equal to $638.42 + 88.09 + 1861.56 = 2588.07$ and the variance is estimated as 258.807 as before.

The regression equation is then:

$$\hat{y} = 74.18 + 2.5\xi_1 - 0.79\xi_2.$$
$$(\pm 6.73) \quad (\pm 1.19) \quad (\pm 0.36)$$

(1.4)

The values predicted by Equation (1.4) are the same as those predicted by Equation (1.3). For example, the value predicted for $t = 1$ by Equation (1.4) is:

$$74.1769 - 2.5110 \times 6 - 0.7934 \times 22 = 41.6561.$$

Values of orthogonal polynomials are available in sets of statistical tables or can be easily calculated from the recurrence relation (1.3) for equally spaced intervals. However, the values must be calculated as needed when the intervals are unequal.

Fuller accounts of the use of matrices in regression analyses are given in Chapter 2 of Draper and Smith (1981) and in Chapter 13 of Steel and Torrie (1980).

1.6 PERIODIC REGRESSION

The cyclic behaviour displayed by many biological variables can often be approximated by periodic regression equations. The independent variable is cyclic in periodic regression. For example, the variable $\cos(2\pi d/365)$, where d is day of the year, is cyclic because its values are repeated every 365 days. The periodic regression equation relating an observed variable y and time t can be written

$$\hat{y} = a_0 + A\cos(\omega t - \theta),$$

(1.5)

where

1. $\hat{y}$ is the predicted or expected value of y;
2. $a_0 = \bar{y}$;
3. $A(>0)$ is the amplitude and is equal to half the range in y from minimum to maximum;
4. ω is a constant that converts t to $360°$ or 2π radians;
5. θ is the (acro) phase angle or the time when the maximum occurs between 0 and 2π.

Equation (1.5) is conveniently written

$$\hat{y} = a_0 + A\cos\theta\cos\omega t + A\sin\theta\sin\omega t$$
$$= b_0 + b_1\cos\omega t + b_2\sin\omega t.$$

(1.6)

Equation (1.6) is linear in $\cos \omega t$ and $\sin \omega t$ and the coefficients b_0, b_1 and b_2 can be estimated as usual. Then

$$\theta = \tan^{-1}(b_2/b_1) \quad \text{and} \quad A = (b_1^2 + b_2^2)^{1/2}.$$

Equation (1.5) can be extended as follows:

$$\hat{y} = a_0 + A_1 \cos(\omega t - \theta_1) + A_2 \cos(2\omega t - \theta_2) + \cdots + A_k \cos(k\omega t - \theta_k).$$

This leads to the following generalization of the regression Equation (1.6):

$$\hat{y} = b_0 + b_{11}\cos \omega t + b_{12}\sin \omega t + b_{21}\cos 2\omega t + b_{22}\sin 2\omega t$$
$$+ \cdots + b_{k1}\cos k\omega t + b_{k2}\sin k\omega t.$$

This regression is a trigonometric polynomial.

The data in Table 1.4 will be used to fit a periodic regression. The values given in column 2 and plotted in Figure 1.3 are transformed values of the

Table 1.4   Transformed numbers of citrus red mites, y, counted on 26 fortnights and variables used to fit a periodic regression to the counts

Time t	Count y	d 0.242t	2d	cos d	sin d	cos 2d	sin 2d	$\hat{y}$
1	1.70	0.242	0.483	0.971	0.239	0.885	0.465	1.68
2	1.38	0.483	0.967	0.885	0.465	0.568	0.823	1.57
3	1.27	0.725	1.450	0.749	0.663	0.121	0.993	1.38
4	0.50	0.967	1.933	0.568	0.823	−0.355	0.935	1.16
5	1.27	1.208	2.417	0.355	0.935	−0.749	0.663	0.95
6	1.00	1.450	2.900	0.121	0.993	−0.971	0.239	0.80
7	1.04	1.692	3.383	−0.121	0.993	−0.971	−0.239	0.74
8	0.64	1.933	3.867	−0.355	0.935	−0.749	−0.663	0.77
9	0.74	2.175	4.350	−0.568	0.823	−0.355	−0.935	0.87
10	0.64	2.417	4.833	−0.749	0.663	0.121	−0.993	1.00
11	1.62	2.658	5.317	−0.885	0.465	0.586	−0.823	1.12
12	1.14	2.900	5.800	−0.971	0.239	0.885	−0.465	1.19
13	1.14	3.142	6.283	−1.000	0.000	1.000	0.000	1.17
14	1.27	3.383	6.766	−0.971	−0.239	0.885	0.465	1.05
15	0.50	3.625	7.250	−0.885	−0.465	0.568	0.823	0.86
16	0.00	3.867	7.733	−0.749	−0.663	0.121	0.993	0.63
17	1.14	4.108	8.216	−0.568	−0.823	−0.355	0.935	0.41
18	0.00	4.350	8.700	−0.355	−0.935	−0.748	0.663	0.24
19	1.00	4.592	9.183	−0.121	−0.993	−0.971	0.239	0.18
20	0.00	4.833	9.666	0.121	−0.993	−0.971	−0.23	0.24
21	0.50	5.075	10.150	0.355	−0.935	−0.749	−0.663	0.42
22	0.00	5.317	10.633	0.568	−0.823	−0.355	−0.935	0.69
23	0.64	5.558	11.116	0.748	−0.663	0.120	−0.993	1.00
24	0.94	5.800	11.600	0.885	−0.465	0.568	−0.823	1.31
25	2.26	6.042	12.083	0.971	−0.239	0.885	−0.465	1.54
26	2.04	6.283	12.566	1.000	0.000	1.000	0.000	1.67

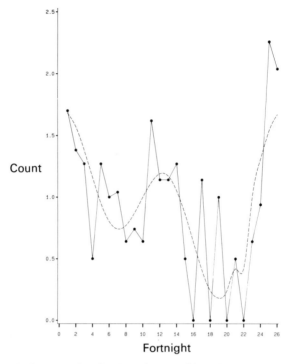

Figure 1.3   Periodic regression fitted to the transformed counts given in Table 1.4.

total numbers of red mites counted on 160 leaves on citrus trees each fortnight for a year. Assuming an annual cycle in the number of mites present, a complete cycle of 2π radians in circular measure is equivalent to 26 fortnights in calendar time, so that ω in the variables $\cos \omega t$, etc., is equal to $2\pi/26 = 0.242$. The values of $d = 0.242t$ are given in column 3 of Table 1.4 and the variables derived from this are given in the columns following it. The periodic regression is fitted by regressing the y-values on the values of $\cos d$, $\sin d$, $\cos 2d$ and $\sin 2d$ as usual. The results are given in Table 1.5; the highly significant term for $\cos 2d$ in the analysis indicates that the second harmonic is needed, although the coefficient of $\sin 2d$ is very small and non-significant.

The fitted equation is:

$$\hat{y} = \underset{(\pm 0.087)}{0.95} + \underset{(\pm 0.124)}{0.25} \cos d + \underset{(\pm 0.122)}{0.28} \sin d + \underset{(\pm 0.121)}{0.47} \cos 2d + \underset{(\pm 0.125)}{0.002} \sin 2d.$$

The values predicted by this equation are given in column 9 of Table 1.4 and were used to plot the curve in Figure 1.3

Periodic regression is treated in detail in Chapter 17, Vol. 2 of Bliss (1970) and in Chapter 8 of Batschelet (1981).

Table 1.5 Analysis of variance and estimates of regression coefficients for the periodic regression of count on time

Source	DF	$MS \times 10^2$	F	P
$\cos d$	1	80.8	4.1	0.056
$\sin d$	1	104.5	5.3	0.032
$\cos 2d$	1	297.3	15.1	0.008
$\sin 2d$	1	000.0	0.0	0.990
Residual	21	19.7		

Variable	Coefficient	Standard error
Constant	0.9477	0.087
$\cos d$	0.2513	0.124
$\sin d$	0.2815	0.124
$\cos 2d$	0.4708	0.124
$\sin 2d$	0.0016	0.124

1.7 SERIAL CORRELATION

Observations taken on an individual on successive occasions or taken on parts of the same unit are often found to be correlated, and the correlation is usually larger the closer the observations are in space or time. This type of correlation is called serial correlation. It must be remembered that the residual mean square does not estimate the variance when least squares is used to fit a regression model to correlated data. In particular, the residual mean square underestimates the variance when neighbouring observations are positively correlated, as would be expected in most biological series.

The occurrence of runs of positive and negative deviations from a fitted trend indicates lack-of-fit for independent observations but may indicate serial correlation in series of obervations.

1.8 ANALYSIS OF VARIANCE

The calculations required for an analysis of variance can be conveniently carried out by regression analysis using linear models, and, in fact, this is the method used in most computer programs. The experiment on live weight of ewes described in detail in Chapter 2 will be used to review the method. Four types of pasture were grazed by the ewes at two stocking rates. The eight treatments were replicated in three randomized complete blocks.

The analysis of variance is given below:

Source	DF
Blocks	2
Pastures (P)	3
Stocking rates (SR)	1
P × SR	3
Error	14

The linear model to be used to obtain this analysis must contain a variable for each degree of freedom except for the degrees of freedom for error. The error sum of squares is estimated by the residual in the analysis. The model is:

$$\hat{y} = c_0 + c_1 x_1 + c_2 x_2 + c_3 x_3 + c_4 x_4 + c_5 x_5 + c_6 x_6 + c_7 x_7 + c_8 x_8 + c_9 x_9.$$

$$\underbrace{\qquad}_{\text{blocks}} \quad \underbrace{\qquad}_{\text{pastures}} \quad \underbrace{\text{stocking rate}} \quad \underbrace{\text{P} \times \text{SR}}$$

(1.7)

As indicated, the variables x_1 and x_2 estimate the effects of blocks and can be called block-variables; x_3, x_4 and x_5 are pasture-variables; x_6 is the stocking rate-variable; and x_7, x_8 and x_9 are the P × SR interaction-variables.

The values of one set of variables that could be used to obtain the analysis are given in Table 1.6. These values form the design matrix for the analysis of variance. The estimates of the coefficients $c_0, \ldots, c_9$ and the variation accounted for by each variable are obtained by regressing live weight on the values of $x_1, \ldots, x_9$. Then, writing SSx_1 for the sum of squares accounted for by the regression on x_1, etc.:

sum of squares for blocks	$= SSx_1 + SSx_2$
sum of squares for pastures	$= SSx_3 + SSx_4 + SSx_5$
sum of squares for stocking rate	$= SSx_6$
sum of squares for P × SR	$= SSx_7 + SSx_8 + SSx_9.$

Variables such as those for pasture and interaction can be defined in a number of ways to suit the aims of the experiment. For example, a different set of variables from those used in this section were considered to be appropriate for the aims of this experiment. These variables are defined in section 2.1 and used for the analysis of variance in section 2.2. The analysis in section 2.2 indicated that some components of the P × SR interaction were significant and a third set of variables was used in section 2.7.2 to examine the effects of pasture at each stocking rate.

Table 1.6 Values of the *x*-variables of Equation (1.7) for the experiment of Chapter 2

Block	Pasture	Stocking rate	x_0	B1 x_1	B2 x_2	P1 x_3	P2 x_4	P3 x_5	SR x_6	P1 × SR x_7	P2 × SR x_8	P3 × SR x_9
1	1	1	1	1	0	1	1	1	1	1	1	1
		2	1	1	0	1	1	1	−1	−1	−1	−1
	2	1	1	1	0	−1	1	1	1	−1	1	1
		2	1	1	0	−1	1	1	−1	1	−1	−1
	3	1	1	1	0	0	−2	1	1	0	−2	1
		2	1	1	0	0	−2	1	−1	0	2	−1
	4	1	1	1	0	0	0	−3	1	0	0	−3
		2	1	1	0	0	0	−3	−1	0	0	3
2	1	1	1	0	1	1	1	1	1	1	1	1
		2	1	0	1	1	1	1	−1	−1	−1	−1
	2	1	1	0	1	−1	1	1	1	−1	1	1
		2	1	0	1	−1	1	1	−1	1	−1	−1
	3	1	1	0	1	0	−2	1	1	0	−2	1
		2	1	0	1	0	−2	1	−1	0	2	−1
	4	1	1	0	1	0	0	−3	1	0	0	−3
		2	1	0	1	0	0	−3	−1	0	0	3
3	1	1	1	−1	−1	1	1	1	1	1	1	1
		2	1	−1	−1	1	1	1	−1	−1	−1	−1
	2	1	1	−1	−1	−1	1	1	1	−1	1	1
		2	1	−1	−1	−1	1	1	−1	1	−1	−1
	3	1	1	−1	−1	0	−2	1	1	0	−2	1
		2	1	−1	−1	0	−2	1	−1	0	2	−1
	4	1	1	−1	−1	0	0	−3	1	0	0	−3
		2	1	−1	−1	0	0	−3	−1	0	0	3

1.9 MULTIVARIATE ANALYSIS OF VARIANCE

Equation 1.7 defines a univariate analysis of variance (Anova) applicable for the 18 observations taken at each weighing. A similar linear model can be used to obtain a multivariate analysis of variance (Manova) for this experiment. The model is then

$$Y = XC,$$

where Y is a matrix of 18 rows and at most 14 columns, X is the same as before and C is a matrix of regression coefficients with the same number of columns as Y. The number of columns in Y is restricted to the number of degrees of freedom for error in the analysis because the error sums of squares and products matrix, E, is singular when this number is exceeded and the analysis cannot be carried out (section 1.12.1). Contrasts ($\leqslant 14$) can be taken among the observations on each unit and analysed to ensure that

E is non-singular. This leads to a repeated measures Manova. Tests of significance are also based on contrasts among the observations in this book; however, a series of Anovas are used instead of a Manova. The same estimates of treatment effects are obtained by a series of Anovas as by the Manova but the tests of significance are simpler in Anova.

1.10 COMPLEX RESPONSES

Most of this book is concerned with the analysis of experiments in which a number of observations are taken on each individual so that the response of each individual is complex (Cox and Hinkley, 1974, section 1.3). These observations are usually taken on the same attribute but this is not necessarily so. Examples used in the book include:

1. live weights of sheep throughout the year;
2. girth of apple trees at spacings that increase systematically;
3. yield of a pasture at successive cuts.

The analyses have two main steps called first and second stage by Cox and Hinkley and first and second order by Batschelet (1981). Batschelet attributed the latter names to Wallraff who worked on animal orientation. The term stage will be used here to avoid confusion with order of polynomials which will also be used frequently.

Stage 1

Summarize the response of each individual by calculating a number of quantities. Such quantities may be:

1. the mean of the observations and linear and quadratic trends in time or space;
2. periodic trends in time;
3. principal components;
4. estimates of the parameters of growth curves;
5. comparisons between components.

Such quantities are referred to as transformed or derived data, first-order data or summary statistics and are purely descriptive.

Stage 2

Analyse the descriptive statistics calculated in Stage 1 to estimate treatment comparisons specified by the design, and to determine the significance of the treatment comparisons. Complex designs will often require more than two stages. Each stage may involve a number of steps.

Some of the examples could be regarded as growth curves and their analysis approached from this point of view. The general approach of calculating statistics that summarize important features of the observations taken on each individual and analysing these to estimate treatment effects has been adopted. Where more advanced methods could be used to advantage this is pointed out. The assistance of an expert should be sought for these.

The complexity of the descriptive stage varies markedly. Often, only simple linear functions of the observations are needed; sometimes, complex estimation procedures that necessitate using computers are required. The procedure followed when the observations are collected usually determines the appropriate method of calculating the descriptive statistics. This procedure is the **within-individuals design**.

Approximating functions are used to describe trends when no functional form is available from theory. This is usually the case in comparative experiments. Algebraic polynomials (section 1.5) such as

$$\hat{y} = b_0 + b_1 x + b_2 x^2 + \cdots + b_{n-1} x^{n-1}$$

are commonly used to smooth a series of observations taken on y at a series of values of x. $\hat{y}$ is the expected response and x is a quantitative variable such as time or concentration. Periodic regression (section 1.6) involving trigonometrical polynomials of the form

$$\hat{y} = b_0 + b_{11}\cos t + b_{12}\sin t + b_{21}\cos 2t + b_{22}\sin 2t + \cdots$$
$$+ b_{r1}\cos rt + b_{r2}\sin rt$$

is commonly used to smooth periodic series of observations taken at time-points t. Both types of polynomials are used in this book. The algebraic polynomials have proved useful for approximating the behaviour of effects induced by the treatments applied in the experiment. The periodic regressions often describe natural cyclic behaviour well. The two types of polynomials can sometimes be combined with benefit.

Plasma thyroxine concentrations in ten sheep on 52 weekly occasions will be used to introduce the basic method. The means are plotted in Figure 1.4. These means appeared in Wallace (1979) and Mr Wallace has kindly supplied the detailed data. The means appear to follow a cyclic pattern. However, the series of observations on a sheep form a time-series, so that the means also form a time-series. Thus the pattern observed may only be random drift due to correlation between the observations. A trend must be fitted to the points and tested for significance before it can be assumed real. Periodic regression will be used to fit the trend.

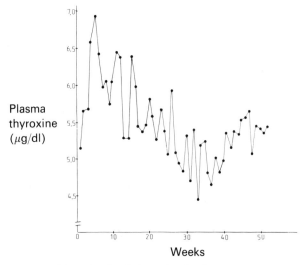

Figure 1.4   Mean levels of thyroxine in sheep on 52 occasions.

1.11 REGRESSION TO BE FITTED TO THE DATA OF FIGURE 1.4

Variation between the ten sheep provides nine degrees of freedom for error, so that at most nine parameters may be estimated to describe the trend in the 52 means (section 1.12.1). The equation to be fitted is

$$\hat{y}=b_0+b_{11}\cos \omega t+b_{12}\sin \omega t+b_{21}\cos 2\omega t+b_{22}\sin 2\omega t$$
$$+b_{31}\cos 3\omega t+b_{32}\sin 3\omega t+b_{41}\cos 4\omega t+b_{42}\sin 4\omega t, \qquad (1.8)$$

where $\omega=2\pi/T$ radians, $T=$ period of the cycle in weeks and $t=$ time in weeks, assuming, $T=52$ weeks, $\omega=2\pi/52=0.12083$ radians.

The regression to be fitted is then

$$\hat{y}=b_0+b_{11}t_{11}+b_{12}t_{12}+b_{21}t_{21}+b_{22}t_{22}+b_{31}t_{31}$$
$$+b_{32}t_{32}+b_{41}t_{41}+b_{42}t_{42}, \qquad (1.9)$$

in familiar notation, with

$$t_{11}=\cos(0.12083t)$$
$$t_{12}=\sin(0.12083t)$$
$$t_{21}=\cos(0.24166t)$$

and so on.

The following steps are used:

1. Calculate the nine coefficients, $b_0, b_{11}, \ldots, b_{42}$, of Equation (1.9) for the data of each sheep.
2. Calculate the mean of each coefficient from the ten estimates and calculate the standard error of each mean; test the significance of the mean coefficients.
3. Estimate which coefficients are required to smooth the data.
4. Carry out a multivariate analysis of variance of the required coefficients to estimate their covariances.
5. Calculate expected values from the polynomial and the variances of these expected values.
6. Use the expected values to draw a smooth curve through the data points.

1.11.1 Fitting Equation (1.9) to individual sheep

The data and the values of $t_{11}, \ldots, t_{42}$ are given in Table 1.7 in the familiar layout for regression analysis. Each column of thyroxine values was regressed on the values of $t_{11}, \ldots, t_{42}$ to obtain the estimates of $b_0, b_{11}, \ldots, b_{42}$ for each sheep. These estimates are given in Table 1.8 with some summary statistics calculated from them.

1.11.2 Fitting Equation (1.9) to the mean thyroxine values

The mean curve is calculated as

$$\hat{y} = \bar{b}_0 + \bar{b}_{11}t_{11} + \bar{b}_{12}t_{12} + \bar{b}_{21}t_{21} + \bar{b}_{22}t_{22} + \bar{b}_{31}t_{31} + \bar{b}_{32}t_{32}$$
$$+ \bar{b}_{41}t_{41} + \bar{b}_{42}t_{42} \tag{1.10}$$

where $\bar{b}_0$ is the mean for b_0, i.e. the overall mean, and the other $\bar{b}$s are the means of the regression coefficients.

The variances of the $\bar{b}$s are calculated as usual, for example,

$$\text{var}(b_0) = s_{b_0}^2 = 1/9(5.29^2 + 4.44^2 + \cdots + 4.54^2 - 54.52^2/10)$$
$$= 0.852037.$$

Then

$$s_{\bar{b}_0}^2 = s_{b_0}^2/10$$
$$= 0.0852037.$$

The significance of $\bar{b}_0$ and of the regression coefficients individually is tested as usual by the ratio $\bar{b}/s_{\bar{b}}$ which follows the t-distribution. The significance of $\bar{b}_0$ and of the regression coefficients jointly is tested with multivariate tests which will be mentioned in section 1.12.1 but which are not usually needed in the procedures used in this book.

Table 1.7 Plasma thyroxine levels in ten sheep on 52 consecutive weeks and values of t variables required to fit Equation (1.9) to them

Week (t)	Sheep 1	2	3	4	5	6	7	8	9	10	t_{11} $\cos\theta$*	t_{12} $\sin\theta$	t_{21} $\cos 2\theta$	t_{22} $\sin 2\theta$	t_{31} $\cos 3\theta$	t_{32} $\sin 3\theta$	t_{41} $\cos 4\theta$	t_{42} $\sin 4\theta$
					Plasma thyroxine ($\mu g/dl$)													
1	4.1	5.8	2.7	6.1	6.3	6.3	3.1	5.7	5.7	6.1	0.9927	0.1205	0.9709	0.2393	0.9350	0.3546	0.8855	0.4647
2	4.7	5.4	3.7	7.3	6.3	7.9	4.1	5.9	5.7	6.1	0.9709	0.2393	0.8856	0.4647	0.7485	0.6631	0.5681	0.8230
3	5.8	5.3	3.7	5.5	6.2	6.1	5.3	6.5	6.2	6.7	0.9350	0.3546	0.7485	0.6631	0.4647	0.8855	0.1205	0.9927
4	6.1	5.8	4.7	6.4	7.8	7.2	5.3	6.9	7.7	7.4	0.8855	0.4647	0.5681	0.8230	0.1205	0.9927	−0.3546	0.9350
5	8.5	6.2	5.1	6.8	7.3	7.5	5.9	5.9	6.9	8.7	0.8230	0.5681	0.3546	0.9350	−0.2393	0.9709	−0.7485	0.6631
6	6.1	5.0	5.5	6.6	9.3	7.9	5.9	6.8	5.2	5.5	0.7485	0.6631	0.1205	0.9927	−0.5681	0.8230	−0.9709	−0.2393
7	5.3	4.5	5.1	6.1	9.7	7.4	5.7	6.3	5.3	4.0	0.6631	0.7485	−0.1205	0.9927	−0.8230	0.5681	−0.9709	−0.2393
8	7.1	4.7	4.9	4.5	9.1	7.1	5.2	5.9	6.5	5.1	0.5681	0.8230	−0.3546	0.9350	−0.9709	0.2393	−0.7485	−0.6631
9	5.8	4.5	4.7	5.9	7.1	7.1	4.5	6.7	5.9	4.3	0.4647	0.8855	−0.5681	0.8230	−0.9927	−0.1205	−0.3546	−0.9350
10	6.0	3.9	5.4	6.1	9.4	8.1	4.8	6.4	5.3	4.5	0.3546	0.9350	−0.7485	0.6631	−0.8855	−0.4647	0.1205	−0.9927
11	5.7	4.1	5.5	6.9	9.3	10.1	5.3	5.7	6.3	5.1	0.2393	0.9709	−0.8855	0.4647	−0.6631	−0.7485	0.5681	−0.8230
12	5.1	4.8	5.3	6.7	8.7	6.5	10.5	6.3	4.8	4.5	0.1205	0.9927	−0.9709	0.2393	−0.3546	−0.9350	0.8855	−0.4647
13	4.1	4.1	5.1	7.4	7.4	5.9	3.7	6.1	5.1	3.7	0	1	−1	0	0	−1	1	0
14	4.5	4.0	5.7	5.6	7.5	5.9	3.7	6.3	4.5	4.7	−0.1205	0.9927	−0.9709	−0.2393	0.3546	−0.9350	0.8855	0.4647
15	5.9	5.3	6.3	6.7	9.5	8.7	4.7	6.1	5.4	4.5	−0.2393	0.9709	−0.8855	−0.4647	0.6631	−0.7485	0.5681	0.8230
16	5.2	4.1	5.4	6.9	9.5	8.2	4.9	6.3	4.5	4.3	−0.3546	0.9350	−0.7485	−0.6631	0.8855	−0.4647	0.1205	0.9927
17	4.9	4.1	4.7	5.7	8.5	8.3	4.6	4.7	4.5	4.0	−0.4647	0.8855	−0.5681	−0.8230	0.9927	−0.1205	−0.3546	0.9350
18	4.3	4.0	5.5	5.6	8.5	6.9	4.1	5.6	4.7	4.1	−0.5681	0.8230	−0.3546	−0.9350	0.9709	0.2393	−0.7485	0.6631
19	5.7	4.2	5.5	5.6	8.3	6.6	4.9	5.7	5.7	4.1	−0.6631	0.7485	−0.1205	−0.9927	0.8230	0.5681	−0.9709	0.2393
20	7.8	4.1	6.3	3.6	6.2	5.3	4.9	6.9	6.3	4.7	−0.7485	0.6631	0.1205	−0.9927	0.5681	0.8230	−0.9709	−0.2393
21	5.5	4.1	6.4	3.9	9.1	5.7	5.3	4.7	5.2	4.5	−0.8230	0.5681	0.3546	−0.9350	0.2393	0.9709	−0.7485	−0.6631
22	6.5	4.5	5.0	3.3	6.1	5.7	4.7	5.3	5.1	5.6	−0.8855	0.4647	0.5681	−0.8230	−0.1205	0.9927	−0.3546	−0.9350
23	5.8	3.5	5.3	3.7	8.9	7.3	6.1	6.3	4.6	4.9	−0.9350	0.3546	0.7485	−0.6631	−0.4647	0.8855	0.1205	−0.9927
24	5.6	4.5	4.7	5.0	5.9	6.9	5.7	6.7	4.9	3.9	−0.9709	0.2393	0.8855	−0.4647	−0.7485	0.6631	0.5681	−0.8230

25	5.5	5.5	3.9	3.5	6.1	6.0	4.6	4.3	3.8	-0.9927	0.1205	0.9709	-0.2393	-0.9350	0.3546	0.8855	-0.4647
26	8.6	6.9	5.1	6.1	5.7	6.7	5.2	4.2	4.7	-1	0	1	0	-1	0	1	0
27	5.6	4.1	3.7	6.5	5.7	8.2	4.4	3.7	3.3	-0.9927	-0.1205	0.9709	0.2393	-0.9350	-0.3546	0.8855	0.4647
28	3.9	4.3	3.9	5.7	5.9	8.2	4.6	3.7	3.6	-0.9709	-0.2393	0.8855	0.4647	-0.7485	-0.6631	0.5681	0.8230
29	4.7	4.3	3.7	3.7	6.3	6.3	4.7	4.3	3.9	-0.9350	-0.3546	0.7485	0.6631	-0.4647	-0.8855	0.1205	0.9927
30	5.5	4.5	4.2	4.9	7.5	7.1	5.5	4.3	3.7	-0.8855	-0.4647	0.5681	0.8230	-0.1205	-0.9927	-0.3546	0.9350
31	3.8	4.1	5.2	4.9	3.4	6.7	5.3	4.4	3.4	-0.8230	-0.5681	0.3546	0.9350	0.2393	-0.9709	-0.7485	0.6631
32	6.1	3.9	5.3	4.9	6.3	7.7	5.7	4.9	3.8	-0.7485	-0.6631	0.1205	0.9927	0.5681	-0.8230	-0.9709	0.2393
33	4.7	3.1	3.9	2.0	7.7	5.5	4.3	3.9	3.5	-0.6631	-0.7485	-0.1205	0.9927	0.8230	-0.5681	-0.9709	-0.2393
34	4.7	6.5	4.5	6.7	5.5	6.0	4.5	3.7	4.3	-0.5681	-0.8230	-0.3546	0.9350	0.9709	-0.2393	-0.7485	-0.6631
35	4.5	5.7	4.6	4.4	8.0	5.7	5.5	4.3	4.0	-0.4647	-0.8854	-0.5681	0.8230	0.9927	0.1205	-0.3546	-0.9350
36	3.9	4.1	6.4	2.7	6.3	6.9	4.1	3.8	3.8	-0.3546	-0.9350	-0.7485	0.6631	0.8855	0.4647	0.1205	-0.9927
37	3.7	3.9	4.5	4.4	7.0	6.4	4.7	3.3	3.3	-0.2393	-0.9709	-0.8855	0.4647	0.6631	0.7485	0.5681	-0.8230
38	4.3	3.5	5.7	4.1	6.6	6.3	5.4	4.1	5.3	-0.1205	-0.9927	-0.9709	0.2393	0.3546	0.9350	0.8855	-0.4647
39	5.0	3.1	5.9	2.7	6.3	6.4	5.0	4.5	3.7	0	-1	-1	0	0	1	1	0
40	5.3	3.8	4.9	3.4	7.5	5.5	4.9	4.1	4.3	0.1205	-0.9927	-0.9709	-0.2393	-0.3546	0.9350	0.8855	0.4647
41	4.3	4.2	5.5	2.8	8.1	7.4	6.0	4.7	4.6	0.2393	-0.9709	-0.8855	-0.4647	-0.6631	0.7485	0.5681	0.8230
42	5.3	4.1	5.3	4.3	6.0	6.5	6.1	4.4	3.3	0.3546	-0.9350	-0.7485	-0.6631	-0.8855	0.4647	0.1205	0.9927
43	4.0	4.5	5.8	4.9	6.1	7.1	5.7	5.1	4.3	0.4647	-0.8855	-0.5681	-0.8230	-0.9927	0.1205	-0.3546	0.9350
44	3.9	3.5	5.7	5.8	5.5	6.9	5.5	4.7	4.7	0.5681	-0.8230	-0.3546	-0.9350	-0.9709	-0.2393	-0.7485	0.6631
45	4.7	4.9	4.9	6.0	6.4	6.9	5.1	4.9	4.3	0.6631	-0.7485	-0.1205	-0.9927	-0.8230	-0.5681	-0.9709	0.2393
46	5.3	4.6	5.5	4.6	5.9	8.1	5.5	5.1	4.7	0.7485	-0.6631	0.1205	-0.9927	-0.5681	-0.8230	-0.9709	-0.2393
47	5.4	4.1	7.1	5.7	7.2	5.9	4.7	5.1	4.7	0.8230	-0.5681	0.3546	-0.9350	-0.2393	-0.9709	-0.7485	-0.6631
48	5.3	3.7	5.2	3.4	5.7	5.8	4.7	4.9	4.8	0.8855	-0.4647	0.5681	-0.8230	0.1205	-0.9927	-0.3546	-0.9350
49	5.3	3.9	6.2	6.1	6.9	6.3	4.5	4.3	4.3	0.9350	-0.3546	0.7485	-0.6631	0.4647	-0.8855	0.1205	-0.9927
50	4.9	3.7	6.2	5.9	6.3	7.4	4.7	4.3	4.1	0.9709	-0.2393	0.8855	-0.4647	0.7485	-0.6631	0.5681	-0.8230
51	5.1	4.1	5.9	5.6	6.2	5.1	4.5	4.7	4.7	0.9927	-0.1206	0.9709	-0.2393	0.9350	-0.3547	0.8855	-0.4647
52	5.7	3.9	5.9	4.8	7.1	6.1	4.8	6.9	4.1	-1	0	1	0	1	0	-1	0

$*\theta = 0.12083t.$

Table 1.8 Estimates of the coefficients of Equation (1.9) for ten sheep and means, variances, standard errors and Student's t-values calculated from these

Sheep	b_0	b_{11} ($\cos\theta$*)	b_{12} ($\sin\theta$)	b_{21} ($\cos 2\theta$)	b_{22} ($\sin 2\theta$)	b_{31} ($\cos 3\theta$)	b_{32} ($\sin 3\theta$)	b_{41} ($\cos 4\theta$)	b_{42} ($\sin 4\theta$)
1	5.290	-0.0238	0.6127	0.4167	0.0773	-0.3410	0.3506	-0.2458	-0.1620
2	4.438	0.0771	0.1968	0.2926	0.3408	-0.0572	0.1656	-0.0258	0.1830
3	5.131	0.2066	0.0150	-0.3745	-0.5459	0.0926	-0.0752	-0.1713	-0.3496
4	5.156	0.5942	0.9452	0.1146	0.2502	-1.0451	-0.5440	0.1239	0.3720
5	7.136	0.1418	1.1219	-0.7647	0.0812	0.0491	0.2160	-0.1802	-0.1461
6	6.840	0.0572	0.4098	-0.1653	0.1963	-0.3731	-0.2318	-0.0264	0.3269
7	5.060	0.0165	0.0371	-0.2464	0.0681	-0.4771	0.0642	-0.1105	-0.0332
8	6.040	0.3837	0.1848	0.2526	-0.1685	-0.2133	-0.1009	-0.0411	-0.0480
9	4.881	0.5588	0.6757	0.0755	0.0756	-0.1204	0.3370	-0.3569	0.1109
10	4.538	0.6869	0.4359	0.3016	0.1969	0.0335	0.5599	-0.1414	0.1957
Total	54.510	2.699	4.6349	-0.0973	0.5720	-1.452	0.7414	-1.1755	0.4496
Mean, $\bar{b}$	5.451	0.2699	0.4635	-0.0097	0.0572	-0.1452	0.0741	-0.1176	0.0450
Variance s^2	0.852037	0.069864	0.140915	0.138329	0.06362	0.03881	0.104302	0.018004	0.053592
Variance $\bar{b}$ ($s^2_{\bar{b}}$)	0.085204	0.006986	0.014092	0.013833	0.006363	0.003881	0.010430	0.001800	0.005359
SE $\bar{b}$ ($s_{\bar{b}}$)	0.2919	0.0836	0.1187	0.1176	0.0798	0.0623	0.1021	0.0424	0.0732
$t = \bar{b}/s_{\bar{b}}$	18.7§	3.23‡	3.90‡	<1	<1	-2.33†	<1	-2.77†	<1

*$\theta = 0.12083t$.
†$P < 0.05$.
‡$P < 0.01$.
§$P < 0.001$.

1.11.3 Coefficients required to smooth the means

The significance of the individual $\bar{b}$s is shown in Table 1.8; $\bar{b}_0$, $\bar{b}_{11}$ and $\bar{b}_{12}$ were all large and highly significant and are clearly needed. The remaining coefficients jointly were not significantly different from zero although $\bar{b}_{31}$ and $\bar{b}_{41}$ were significant at $P=0.05$ individually. The fitted equation was

$$\hat{y}=5.45+0.270t_{11}+0.464t_{12}-0.010t_{21}+0.057t_{22}-0.145t_{31}$$
$$+0.074t_{32}-0.118t_{41}+0.045t_{42}. \tag{1.11}$$

The curves obtained from Equation (1.11) and from the reduced equation

$$\hat{y}=5.45+0.270t_{11}+0.464t_{12} \tag{1.12}$$

are shown in Figure 1.5. The reduced Equation (1.12) appears to summarize the main trend in the means. It will be used in section 1.12 to estimate expected values and the times when plasma thyroxine was at its maximum and minimum. The coefficients of Equation (1.12) were obtained from Equation (1.11) without refitting the regression because the columns of t-variables in Table 1.7 were orthogonal to one another.

Procedures for determining model adequacy when fitting polynomial equations to sequences of observations were considered by Roberts and Raison (1986). They recommended using the equation determined by the lowest-order significant coefficient ($P=0.05$) that was followed by two successive non-significant ones according to separate t-tests. This procedure selected the correct polynomial in $95.10 \pm 0.10\%$ of analyses using simulated data. This approach cannot be used rigorously in all the examples to be presented because a number of the regression equations are not simple

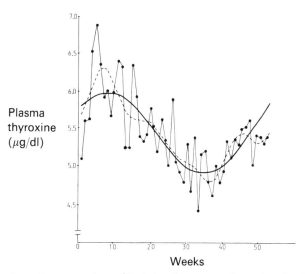

Figure 1.5 Periodic regressions fitted to the mean thyroxine levels of sheep: Equation (1.11) ––––, Equation (1.12) —.

polynomials. Selecting equations on the basis of the highest-order significant coefficient leads to overfitting, in that some high-order coefficients that are not really needed are included in the equation. However, multivariate tests of all the higher-order coefficients together may not detect important coefficients. Although overfitting is undesirable, it does not seriously affect the conclusions that can be drawn from the data, and t-tests are used in preference to the multivariate tests here.

1.12 ESTIMATION FROM THE REGRESSION EQUATION

1.12.1 Multivariate analysis of variance of b_0, b_{11} and b_{21}

The variances of any quantities determined from Equation (1.12) will involve estimates of the covariances among b_0, b_{11} and b_{12} as well as estimates of their variances. The variances have been obtained simply in Table 1.8. The covariances can also be obtained easily in this simple example as the sum of products of the individual values in Table 1.8. For example, the covariance of $\bar{b}_0$ and $\bar{b}_1$ can be calculated as follows:

$$\text{cov}(b_0 b_1) = \tfrac{1}{9}[(5.290 \times -0.0238 + 4.438 \times 0.0771 + \cdots + 4.538 \times 0.6869)$$
$$- 54.51 \times 2.699/10)]$$

$$= -0.72340/9,$$

so that

$$\text{cov}(\bar{b}_0 \bar{b}_1) = -0.72340/(9 \times 10) = -0.0080378.$$

However, the variances and covariances are usually obtained in a multivariate analysis of quantities such as b_0, b_{11} and b_{12}. The calculations are extensions of the analysis of covariance. The squares and products shown in Table 1.9 are calculated, and are partitioned according to the experimental design. The appropriate analysis for this simple design is given in Table 1.10.

To obtain this analysis the entries in the total line in Table 1.9 were copied to the total line in Table 1.10 and the values in the mean line of Table 1.10 were calculated from the mean line of Table 1.8 as 10×5.451^2, 10×0.2669^2, 10×0.4635^2, $10 \times 5.451 \times 0.2669$, $10 \times 5.451 \times 0.4635$ and $10 \times 0.2669 \times 0.4635$. Then, as usual, the values for the error line were calculated by finding the difference between the values in the total and mean lines. To obtain the analysis by a multivariate regression program, the values of b_0, b_{11} and b_{12} in Table 1.8 would be regressed on a column of ones. This is referred to as fitting the mean; the mean is included automatically in most analyses.

The matrix composed of the sums of squares and products from the error line is usually called E. That associated with the mean or treatments is

Table 1.9 Sums of squares and products of b_0, b_{11} and b_{12} of Table 1.8

Sheep	b_0^2	b_{11}^2	b_{12}^2	$b_0 b_{11}$	$b_0 b_{12}$	$b_{11} b_{12}$
1	27.984100	0.00056644	0.37540129	−0.1259020	3.2411830	−0.01458226
2	19.695844	0.00594441	0.03873024	0.3421698	0.8733984	0.01517328
3	26.327161	0.04268356	0.00022500	1.0600646	0.0769650	0.00309900
4	26.584336	0.35307364	0.89340304	3.0636952	4.8734512	0.56163784
5	50.922496	0.02010724	1.25865961	1.0118848	8.0058784	0.15908542
6	46.785600	0.00327184	0.16793604	0.3912480	2.8030320	0.02344056
7	25.603600	0.00027225	0.00137641	0.0834900	0.1877260	0.00061215
8	36.481600	0.14722569	0.03415104	2.3175480	1.1161920	0.07090776
9	23.824161	0.31225744	0.45657049	2.7275028	3.2980917	0.37758116
10	20.593444	0.47183161	0.19000881	3.1171522	1.9781142	0.29941971
Total	304.802342	1.35723412	3.41646197	13.9888534	26.4540319	1.49637462

Table 1.10 Partitioning of the sums of squares and products of b_0, b_{11} and b_{12}

Source of variation	DF	Sums of square			Sums of products		
		b_0^2	b_{11}^2	b_{12}^2	$b_0 b_{11}$	$b_0 b_{12}$	$b_{11} b_{12}$
H mean	1	297.13401	0.7284601	2.14822980	14.712249	25.2648399	1.25094951
E error (between sheep)	9	7.668332	0.62877402	1.26823217	−0.7233956	1.1891920	0.24541511
T total	10	304.802342	1.35723412	3.41646197	13.9888534	26.4540319	1.49637462

called **H**. Here,

$$E = \begin{bmatrix} 7.668332 & -0.7233956 & 1.1891920 \\ -0.7233956 & 0.62877402 & 0.24541511 \\ 1.1891920 & 0.24541511 & 1.26823217 \end{bmatrix}.$$

S_b, the error variance–covariance matrix of b_0, b_{11} and b_{12} equals **E** divided by nine, the number of degrees of freedom for error. Here,

$$S_b = \begin{bmatrix} 0.852037 & & \\ -0.080377 & 0.069864 & \\ 0.132132 & 0.027268 & 0.140915 \end{bmatrix} \tag{1.13}$$

$$= \begin{bmatrix} s_{00} & & \\ s_{01} & s_{11} & \\ s_{02} & s_{12} & s_{22} \end{bmatrix}.$$

The error correlation matrix, **R**, was then

$$R = \begin{bmatrix} 1 & & \\ -0.331055 & 1 & \\ 0.382599 & 0.274644 & 1 \end{bmatrix}$$

$$= \begin{bmatrix} 1 & & \\ r_{01} & 1 & \\ r_{02} & r_{12} & 1 \end{bmatrix},$$

$$r_{01} = -0.080377/(0.852037 \times 0.069864)^{1/2}$$
$$r_{02} = 0.132132/(0.852037 \times 0.140915)^{1/2}$$
$$r_{12} = 0.027268/(0.069864 \times 0.140915)^{1/2}.$$

The significance of $\bar{b}_0$, $\bar{b}_{11}$ and $\bar{b}_{12}$ jointly can be tested by the ratio

$$\Lambda = |E|/|H + E|$$

where $|E|$ stands for the determinant of **E**.

Here,

$$\Lambda = 0.0167196$$

for which

$$P < 0.001.$$

$|E| = 0$ when the number of coefficients exceeds the number of degrees of freedom for error. **E** is then said to be singular and it is impossible to define a unique statistical distribution for the coefficients. In this example, there are nine degrees of freedom for error. Therefore, at most nine coefficients can be used to describe the trends in the values of thyroxine when this testing procedure is used.

1.12.2 Expected (fitted) values

Equation (1.12),

$$\hat{y} = 5.45 + 0.270 t_{11} + 0.464 t_{12},$$

was used to calculate the values of y corresponding to the observed t_{11} and t_{12}. These expected values were used to plot this curve in Figure 1.5. The variances of the expected values are given by

$$\text{var}(\hat{y}) = s_{00} + t_{11}^2 s_{11} + t_{12}^2 s_{22} + 2 t_{11} s_{01} + 2 t_{12} s_{02}$$
$$+ 2 t_{11} t_{12} s_{12}$$

where the s_{ij} are the elements of $S_{\bar{b}} = S_b/10$; S_b is given by Equation (1.13).

1.12.3 Maximum and minimum values

In the original time-scale of weeks, Equation (1.12) becomes

$$\hat{y} = 5.45 + 0.270 \cos(0.12083t) + 0.464 \sin(0.12083t) \qquad (1.14)$$

The times when such a periodic curve reaches its maximum and minimum values provide examples of the estimates that can be obtained from a regression equation. These values are obtained by differentiating Equation (1.14) with respect to t and equating the result to zero. The simple equation obtained in this example has the following analytic solutions:

$$t_{\max} = \tan^{-1}(0.464/0.270)/0.12083$$
$$= 1.0437/0.12083$$
$$= 8.64 \text{ weeks} \qquad (1.15)$$

and

$$t_{\min} = (\pi + 1.0437)/0.12083$$
$$= 34.64 \text{ weeks.}$$

The variance of $t_{\max}$ (or $t_{\min}$) can be estimated by writing Equation (1.15) as

$$t_{\max} = \tan^{-1}(\bar{b}_{12}/\bar{b}_{11})/0.12083 \qquad (1.16)$$

and applying the formula for the variance of a function of random variables (Stuart and Ord, 1987, section 10.5). This gives

$$\mathrm{var}(t_{\max}) \doteq (\bar{b}_{12}^2 \mathrm{var}\, \bar{b}_{11} - 2\bar{b}_{11}\bar{b}_{12} \mathrm{cov}\, \bar{b}_{11}\bar{b}_{12} + \bar{b}_{11}^2 \mathrm{var}\, \bar{b}_{12})/0.12083^2(\bar{b}_{11}^2 + \bar{b}_{12}^2)^2$$
$$= (\bar{b}_{12}^2 s_{11} - 2\bar{b}_{11}\bar{b}_{12} s_{12} + \bar{b}_{11}^2 s_{22})/0.12083^2(\bar{b}_{11}^2 + \bar{b}_{12}^2)^2$$
$$= 1.512724,$$

so that standard error of $t_{\max} = \pm 1.23$ approximately.

However, the average $t_{\max}$ and its variance can be calculated directly by calculating for each sheep and calculating the mean and variance of these. It is convenient to start with the values of $0.12083 t_{\max}$. From Equation (1.16)

$$0.12083 t_{\max} = \tan^{-1}(b_{12}/b_{11}).$$

The values of b_{11} and b_{12} in Table 1.8 give the following estimates of $0.12083 t_{\max}$ for the 10 sheep:

1.532, 1.197, 0.072, 1.010, 1.445, 1.432, 1.152, 0.449, 0.880, 0.565.

Then

$$0.12083 \bar{t}_{\max} = 0.9734,$$

so

$$\bar{t}_{\max} = 8.06 \text{ weeks},$$

and

$$\mathrm{var}(0.12083 t_{\max}) = 0.23234,$$

so

$$\mathrm{var}(t_{\max}) = 0.23234/0.12083^2$$
$$= 15.9138.$$

Therefore

$$\mathrm{var}(\bar{t}_{\max}) = 1.59138$$

and

$$\text{standard error } (\bar{t}_{\max}) = \pm 1.26.$$

Thus, complicated or approximate expressions for the variances of estimates may often be avoided by calculating the values of the estimator for each individual and estimating the variance from them.

1.13 DISCUSSION

Without doubt the procedure used in section 1.11 to fit the regression to the mean thyroxine values is roundabout and difficult to follow at first. The regression could have been obtained easily by regressing the mean values directly on the t-variables of Table 1.7. However, this does not provide the correct variances because the deviations from regression do not estimate the variance when the obervations are correlated.

The correct variances are automatically produced in the indirect method of fitting. Direct fitting is considered in more detail in Chapter 7.

Equation (1.17) was used in section 1.12.3 to estimate t_{max}:

$$\hat{y} = \bar{b}_0 + \bar{b}_{11}t_{11} + \bar{b}_{12}t_{12} \tag{1.17}$$

This is a reduced version of Equation (1.9). It was chosen for simplicity so that the method could be followed in detail. The maxima and minima are given by the roots of the equation $d\hat{y}/dt = 0$. For Equation (1.17) this has the analytic solution used in section 1.12.3. If Equation (1.9) were used, the maxima and minima would be estimated numerically because then the equation $d\hat{y}/dt = 0$ does not have an analytic solution. This is the only complication introduced. For example, the absolute maximum of Equation (1.11) was estimated numerically for each sheep using the Newton–Raphson method to determine the roots of the equation $d\hat{y}/dt = 0$. The values obtained were:

6.72, 4.15, 11.31, 9.88, 9.75, 12.00, 8.48, 7.65, 5.71, 4.11

for sheep 1 to sheep 10, respectively. The mean of these values is 7.976 and their variance is 7.860. Therefore, $\bar{t}_{max}$ is estimated to be 7.98 ± 0.887 weeks.

2 A simple factorial grazing experiment measured on 27 occasions

2.1 INTRODUCTION

FitzGerald (1979) described an experiment comparing the performance of ewes grazing four pastures:

1. S – subterranean clover;
2. SP – subterranean clover and phalaris;
3. SPL – subterranean clover, phalaris and lucerne;
4. SL – subterranean clover and lucerne.

These pastures were grazed at 8.9 and 13.3 ewes per hectare. There were three replicates laid out in complete blocks. All pastures were sown in July 1968 and were grazed occasionally until August 1969 when five test ewes were put on each plot.

The live weights of the ewes were recorded on 27 occasions as follows:

1970: five occasions, 29 April, 28 May, 23 July, 17 September, 27 November.

1971: nine occasions, 10 February, 25 February, 22 April, 19 May, 28 June, 19 August, 17 September, 26 October, 1 December.

1972: nine occasions, 25 January, 15 March, 26 April, 15 May, 13 June, 3 July, 7 November, 20 November, 26 December.

1973: four occasions, 13 February, 13 March, 10 April, 28 April.

The experiment was conducted at Wagga Wagga in the southern wheat-growing area of New South Wales. There, the pasture ley mostly consists of subterranean clover with one or more species of annual grass. There are large seasonal fluctuations in pasture growth; in summer usually only dry residues are available and sheep lose weight. This can continue into autumn, depending on the timing of the rain. Clovers and grasses germinate after the autumn rains and make steady slow growth during winter. They grow rapidly in spring and early summer when conditions are usually very good.

The main aims of the experiment were to see if lucerne could provide sufficient feed in summer and autumn to maintain live weight in this region and to compare pastures containing phalaris and subterranean clover with those containing only subterranean clover.

The data are given in Table 2.1 and the means over replicates have been plotted against time in Figure 2.1. Trends in the mean live weight with time are clearly displayed in Figure 2.1, but most of the effects of the treatments are difficult to see. The effect of stocking rate is obvious, but the effects of the pastures are not. The effects of the pastures can also be displayed by using comparisons among the pastures.

The eight treatments provided the following seven comparisons.

1. SR: refers to the difference between ewes at 8.9 per hectare and those at 13.3, calculated as the 8.9 value minus the 13.3 value.
2. Pl (SP−S): the difference in live weight between ewes on SP and those on S, calculated as the SP value minus the S value.
3. P2 (SPL–SL): the difference between ewes on SPL and those on SL.
4. P3 (SPL+SL)−(SP+S): the difference between ewes on lucerne and those not on lucerne.
5. P1 × SR: the interaction of P1 and SR regarded as the difference between P1 at 8.9 ewes per hectare and P1 at 13.3 ewes per hectare.
6. P2 × SR: the interaction of P2 and SR.
7. P3 × SR: the interaction of P3 and SR.

The live weights at the first weighing time will be used to illustrate these comparisons. The treatment totals in kilograms over the three replicates are conveniently set out in Table 2.2. Then

$$SR = \text{mean live weight of ewes at 8.9 per hectare}$$
$$- \text{mean live weight of ewes at 13.3 per hectare}$$
$$= 71.775 - 69.142$$
$$= 2.633 \text{ kg},$$

$$P1 = SP - S$$
$$= 69.433 - 64.617$$
$$= 4.816,$$

$$P2 = SPL - SL$$
$$= 72.750 - 75.033$$
$$= -2.283,$$

$$P3 = (SL + SPL) - (S + SP)$$
$$= (72.750 + 75.033)/2 - (64.617 + 69.433)/2$$
$$= 6.8665,$$

$$P1 \times SR = P1 \text{ (8.9 ewes/ha)} - P1 \text{ (13.3 ewes/ha)}$$
$$= 1/3[(212.6 - 195.0) - (204.0 - 192.7)]$$
$$= 2.10,$$

Table 2.1 Live weight of ewes (hectograms) grazing four pastures (P) at two rates of stocking (SR) in three blocks (B) on 27 dates

BP	SR	1	2	3	4	5	6	7	8	9	10	11	12	13	14	15	16	17	18	19	20	21	22	23	24	25	26	27
11	1	657	635	679	718	801	678	726	724	714	734	765	800	856	798	810	850	788	808	788	722	808	804	766	637	647	632	623
21	1	649	626	666	698	777	756	683	697	685	710	792	792	848	813	817	864	770	787	749	715	845	836	812	656	656	640	625
31	1	644	611	655	707	813	668	688	674	671	695	738	771	835	760	760	808	726	738	695	528	773	762	724	583	586	576	556
12	1	692	658	712	719	819	719	743	771	739	760	832	866	901	851	877	899	837	837	745	656	773	779	757	616	606	583	576
22	1	685	645	679	741	837	823	768	728	723	731	768	807	868	805	821	868	779	778	746	684	792	746	720	566	563	540	534
32	1	749	691	733	799	909	768	768	746	614	756	798	833	900	820	842	919	847	858	839	747	838	834	798	645	642	925	613
11	2	624	586	606	660	775	647	657	642	613	610	651	682	790	746	750	830	727	716	707	561	811	796	714	470	572	570	550
21	2	687	625	641	673	786	711	704	676	541	650	675	546	811	759	759	810	733	754	700	608	732	726	678	560	565	552	544
31	2	616	569	591	639	725	610	615	632	618	639	652	680	750	702	714	741	694	714	683	606	691	673	630	396	418	535	501
12	2	650	608	624	677	612	750	671	657	652	653	668	714	791	722	723	803	716	702	651	579	706	693	644	406	534	519	516
22	2	723	661	663	722	824	699	705	708	670	676	700	730	789	745	747	812	742	726	678	591	698	695	646	537	563	544	535
32	2	667	599	606	669	789	657	673	672	643	649	679	715	798	763	751	815	740	720	670	568	710	704	634	474	507	528	509
13	1	756	694	708	759	833	780	805	750	727	746	754	796	866	839	874	902	794	767	745	680	789	789	771	618	621	612	600
23	1	708	660	680	727	798	726	761	730	706	732	755	787	821	788	816	818	739	593	702	668	780	787	745	596	586	594	610
33	1	737	692	711	755	792	772	814	778	761	786	803	818	856	847	873	883	769	749	737	692	829	805	790	636	662	656	659
14	1	809	737	754	797	870	878	888	839	810	808	785	830	880	875	927	926	815	809	783	732	861	855	822	661	690	689	678
24	1	779	716	723	782	881	820	852	818	790	808	818	835	909	895	957	950	851	839	810	752	869	859	825	651	665	658	659
34	1	748	693	726	781	850	800	816	793	786	797	830	853	901	881	895	903	807	754	719	627	813	813	789	635	654	654	645
13	2	687	573	600	650	736	654	692	638	622	628	725	784	846	818	873	888	756	672	703	624	753	740	693	470	588	536	530
23	2	747	651	648	706	807	754	778	721	714	709	684	737	811	815	875	881	762	748	707	624	742	742	694	556	559	548	550
33	2	730	647	630	676	796	740	752	712	694	694	685	724	794	795	852	862	748	731	706	626	732	730	672	535	516	518	489
14	2	696	670	657	694	772	824	760	735	596	719	681	707	770	764	804	816	705	691	633	562	675	670	627	437	528	544	526
24	2	736	633	615	678	759	674	728	653	641	629	602	628	733	751	807	827	696	675	633	550	685	662	630	486	480	481	519
34	2	734	640	639	696	766	626	764	686	681	677	683	722	801	805	822	844	743	724	695	617	736	734	695	564	546	551	538

Columns 1–27 are *Weighing times*.

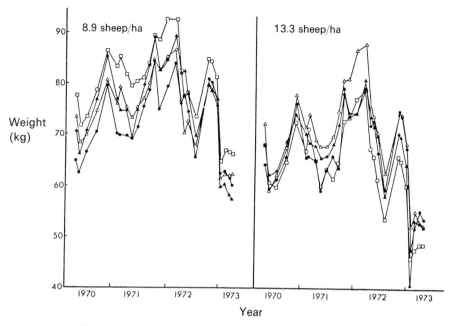

Figure 2.1   Live weights on 27 occasions for ewes grazing four pasture types at two stocking rates:
- ● subterranean clover;
- ▼ subterranean clover and phalaris;
- ▽ lucerne, subterranean clover and phalaris;
- □ lucerne and subterranean clover.

Table 2.2

Stocking rate (ewes per hectare)	Pasture				Mean per ewe
	S	SP	SPL	SL	
8.9	195.0	212.6	220.1	233.6	71.775
13.3	192.7	204.0	216.4	216.6	69.142
Mean per ewe	64.617	69.443	72.750	75.033	

$$P2 \times SR = P2 \ (8.9 \ ewes/ha) - P2 \ (13.3 \ ewes/ha)$$
$$= 1/3[(220.1 - 233.6) - (216.4 - 216.6)]$$
$$= -4.433,$$

$$P3 \times SR = P3 \ (8.9 \ ewes/ha) - P3 \ (13.3 \ ewes/ha)$$
$$= 1/6[(220.1 + 233.6) - (195.0 + 212.6)]$$
$$- 1/6[(216.4 + 216.6) - (192.7 + 204.0)]$$
$$= 1.663.$$

The values of each of these comparisons are estimated at each weighing and the 27 values of each are regressed on time. This amounts to seven regression analyses but the regressions are fitted indirectly. A number of steps can be distinguished; these are summarized below and then dealt with in detail.

Preliminary stage

2.2 Estimate the values of the treatment comparisons for each weighing time. This is conveniently done by regressing the live weights at each weighing on variables describing these comparisons (*x*-variables).

2.3 Plot the values of the comparisons against time to help select appropriate regression equations.

2.4 Choose the regression equation to be used to approximate the trends in the comparisons with time. Assume the regression equation has coefficients $b_0, b_1, \ldots, b_k$.

Stage 1

2.5 Fit the equation chosen in 2.4 to the data for each plot of the experiment to estimate the coefficients $b_0, b_1, \ldots, b_k$ for *each plot*.

Stage 2

2.6 Regress the values of each of the b_k on the *x*-variables of 2.2. The $k+1$ analyses carried out in this step fit the regression equation of 2.4 to the values of the seven treatment comparisons.

2.7 Choose the equations that adequately approximate the trends in the comparisons.

Steps 2.4, 2.5 and 2.6 may need to be repeated a number of times with different regression equations before equations are selected for each treatment comparison.

2.8 Calculate the expected values from the equations chosen in 2.7 and draw the regression lines.

2.2 ESTIMATES OF TREATMENT COMPARISONS

The estimates of the treatment comparisons at each weighing, or multiples of them, are conveniently obtained as the coefficients $c_3, c_4, \ldots, c_9$ of the following regression equation:

$$\hat{y}=c_0x_0+c_1x_1+c_2x_2+c_3x_3+c_4x_4+c_5x_5+c_6x_6+c_7x_7+c_8x_8+c_9x_9 \quad (2.1)$$

where $\hat{y}$ is expected live weight; x_1 and x_2 are block variables, and $x_3,\ldots,x_9$ are treatment variables; c_0, the constant term, is equal to the grand mean here because the design is balanced; $c_1, c_2,\ldots,c_9$ are coefficients relating live weight and the design variables.

Equation (2.1) was fitted to the live weights at each weighing. The design matrix, X is listed in Table 2.3. The matrix was built up column by column by assigning the following values.

1. $x_0 = 1$.
2. $x_1 = 1$ if the observation comes from block 1,
   $ = 0$ if the observation comes from block 2,
   $ = -1$ if the observation comes from block 3.

Table 2.3   Design matrix used to fit Equation (2.1)

Block	P	SR	Mean x_0	B1 x_1	B2 x_2	P1 x_3	P2 x_4	P3 x_5	SR x_6	P1×SR x_7	P2×SR x_8	P3×SR x_9
1	S	8.9	1	1		−1		−1	1	−1		−1
	S	13.3	1	1		−1		−1	−1	1		1
	SP	8.9	1	1		1		−1	1	1		−1
	SP	13.3	1	1		1		−1	−1	−1		1
	SPL	8.9	1	1			1	1	1		1	1
	SPL	13.3	1	1			1	1	−1		−1	−1
	SL	8.9	1	1			−1	1	1		−1	1
	SL	13.3	1	1			−1	1	−1		1	−1
2	S	8.9	1		1	−1		−1	1	−1		−1
	S	13.3	1		1	−1		−1	−1	1		1
	SP	8.9	1		1	1		−1	1	1		−1
	SP	13.3	1		1	1		−1	−1	−1		1
	SPL	8.9	1		1		1	1	1		1	1
	SPL	13.3	1		1		1	1	−1		−1	−1
	SL	8.9	1		1		−1	1	1		−1	1
	SL	13.3	1		1		−1	1	−1		1	−1
3	S	8.9	1	−1	−1	−1		−1	1	−1		−1
	S	13.3	1	−1	−1	−1		−1	−1	1		1
	SP	8.9	1	−1	−1	1		−1	1	1		−1
	SP	13.3	1	−1	−1	1		−1	−1	−1		1
	SPL	8.9	1	−1	−1		1	1	1		1	1
	SPL	13.3	1	−1	−1		1	1	−1		−1	−1
	SL	8.9	1	−1	−1		−1	1	1		−1	1
	SL	13.3	1	−1	−1		−1	1	−1		1	−1
Sum of squares			24	16	16	12	12	24	24	12	12	24

3. $x_2 =$ 0 for block 1,
   $=$ 1 for block 2,
   $= -1$ for block 3.
4. $x_3 = -1$ for pasture S,
   $=$ 1 for pasture SP,
   $=$ 0 for pastures SPL and SL.
5. $x_4 =$ 1 for pasture SPL,
   $= -1$ for SL,
   $=$ 0 for S and SP.
6. $x_5 = -1$ for pastures S and SP,
   $=$ 1 for SPL and SL.
7. $x_6 =$ 1 for 8.9 ewes/ha,
   $= -1$ for 13 ewes/ha.

The values for the interaction variables can be derived similarly. However, they are more easily obtained by multiplying the appropriate elements in the columns that have already been formed.

The relations between the values of the treatment comparisons defined in section 2.1 and the regression coefficients of Equation (2.1) can be obtained from the live weights estimated from Equation (2.1). For example, the estimated live weight at 8.9 ewes per hectare is obtained by averaging the estimates for all pastures grazed at 8.9 ewes per hectare. The estimates from Equation (2.1) and the design matrix are given in Table 2.4.

Similarly, the estimated live weight at 13.3 ewes per hectare is found to be $c_0 - c_6$. The difference is $2c_6$. Because this must equal the value of this comparison calculated in section 2.1, the value of c_6 in this method of fitting is half the value of the comparison calculated directly.

Similarly,

$$P1 = 2c_3,$$
$$P2 = 2c_4,$$
$$P3 = 2c_5;$$

Table 2.4

Pasture		Expected live weight	
S	$c_0 - c_3$	$-c_5 + c_6 - c_7$	$-c_9$
SP	$c_0 + c_3$	$-c_5 + c_6 + c_7$	$-c_9$
SL	c_0	$-c_4 + c_5 + c_6$	$-c_8 + c_9$
SPL	c_0	$+c_4 + c_5 + c_6$	$+c_8 + c_9$
Total	$4c_0$	$+4c_6$	
Average		$c_0 + c_6$	

and

$$P1 \times SR = 4c_7,$$
$$P2 \times SR = 4c_8,$$
$$P3 \times SR = 4c_9.$$

The notation of matrices provides a convenient shorthand for the row by column arrays characteristic of experiments with sequential observations. For example, the data of Table 2.1 form a 24×27 data matrix and the values of the x-variables in Table 2.3 form a 24×10 design matrix.

The 24 equations represented by Equation (2.1) can be rewritten, using matrices, as

$$\hat{Y} = X\Gamma. \tag{2.1a}$$

The estimates of the coefficients are

$$\hat{\Gamma} = (X^TX)^{-1}X^TY \tag{2.2}$$

and the variances and covariances of the estimates are given by

$$V(\hat{\Gamma}) = (X^TX)^{-1}s^2,$$

where s^2 is the error mean square. $(X^TX)^{-1}$ is given in Table 2.5. The values of c_0 and $c_3, \ldots, c_9$ estimated for the 27 weighings are given in Table 2.6.

Table 2.5 $(X^TX)^{-1}$ of Equation (2.2)

	Mean	B1	B2	P1	P2	P3	SR	P1 × SR	P2 × SR	P3 × SR
Mean	1/24	0	0	0	0	0	0	0	0	0
B1	0	1/12	−1/24	0	0	0	0	0	0	0
B2	0	−1/24	1/12	0	0	0	0	0	0	0
P1	0	0	0	1/12	0	0	0	0	0	0
P2	0	0	0	0	1/12	0	0	0	0	0
P3	0	0	0	0	0	1/24	0	0	0	0
SR	0	0	0	0	0	0	1/24	0	0	0
P1 × SR	0	0	0	0	0	0	0	1/12	0	0
P2 × SR	0	0	0	0	0	0	0	0	1/12	0
P3 × SR	0	0	0	0	0	0	0	0	0	1/12

Table 2.6 Estimates of c_0, c_3, c_4, c_5, c_6, c_7, c_8 and c_9 of Equation (2.1) for 27 weighings

Variable and coefficient		1	2	3	4	5	6
Constant term	c_0	70.4583	64.6667	66.4417	71.3458	79.6958	73.0583
P1	c_3	2.4083	1.7500	1.4917	1.9333	0.9417	2.8833
P2	c_4	−1.1417	−1.4333	−1.1417	−1.2917	−1.1333	−1.6333
P3	c_5	3.4333	2.0500	0.9833	1.1625	0.8042	2.3417
Stocking Rate	c_6	1.3167	2.4833	3.7750	3.5125	3.4708	3.5083
P1 × SR	c_7	0.5250	0.2833	0.5750	0.3333	1.9583	0.5833
P2 × SR	c_8	−1.1083	−0.2333	−0.5917	−0.6917	−1.8333	−2.0333
P3 × SR	c_9	0.4083	0.6667	0.5000	0.6625	−0.2375	0.6917

Table 2.6 Cont.

Variable and coefficient		15	16	17	18	19	20
Constant term	c_0	82.2750	85.4958	76.1833	74.4417	71.9125	64.2250
P1	c_3	1.2583	1.7750	1.8583	0.8667	0.0583	0.7083
P2	c_4	−0.4083	−0.2667	−0.4083	−0.7667	−0.0667	−0.1250
P3	c_5	4.1833	2.0042	0.3583	−1.7083	−0.1792	1.1833
Stocking Rate	c_6	3.3000	2.7542	3.1667	2.5167	3.8625	4.9000
P1 × SR	c_7	1.2917	0.9583	1.1250	1.4667	1.5750	1.3250
P2 × SR	c_8	−3.1917	−2.6667	−2.4417	−2.7500	−2.6500	−2.5917
P3 × SR	c_9	−0.7250	−0.5542	−0.1250	−1.4333	−0.0792	0.4083

2.3 PLOTS OF THE VALUES OF $c_0, c_3, \ldots, c_9$ AGAINST TIME

The values of c_0, $c_3, \ldots, c_9$ are plotted against time in Figure 2.2. These graphs illustrate the effects of the treatments much more clearly than those in Figure 2.1.

2.4 REGRESSION OF $c_3, \ldots, c_9$ ON TIME

The experimental design of eight treatments in three replicates provided 14 degrees of freedom for error. This allowed 14 coefficients to be used in the within-individual plot regressions. Hopefully, far fewer than 14 will be needed. Quadratic equations are often good approximations to trends in treatment contrasts with time but a fourth-order polynomial was tried to allow for additional variation.

Weighing time							
7	8	9	10	11	12	13	14
74.0542	71.5833	68.3792	70.8167	72.8708	75.6542	83.0208	79.2875
1.8083	1.9750	1.6583	1.5583	1.7167	3.2833	1.3083	2.1333
−1.7167	−1.6250	−0.6667	−1.1917	0.0583	0.5917	0.0000	−0.5750
4.3625	2.1917	2.6875	1.9583	0.5042	1.1875	0.2125	2.9875
3.2292	3.8167	4.3375	4.7083	5.4958	5.9125	3.9875	2.7542
0.5917	0.5250	−1.5583	0.2417	0.5667	−0.9000	0.8583	1.7333
−1.2167	−1.5750	−2.5333	−1.2917	−2.0750	−2.5417	−2.4500	−2.3750
0.6208	0.8750	0.9292	0.4667	0.2125	−0.7708	−0.0042	0.3875

Weighing time						
21	22	23	24	25	26	27
76.8417	75.9792	71.9833	55.7958	58.1417	57.8708	57.0208
−1.1917	−1.2167	−1.0417	−0.4833	−0.2417	−1.3500	−0.9667
−0.1250	−0.0083	−0.1917	−0.1917	−0.2583	−0.9417	−1.0583
0.3667	0.5792	0.9583	1.2458	0.9833	0.8042	1.3375
4.5833	4.6042	5.6750	6.7042	5.0083	4.3208	4.4625
0.8083	0.5000	0.5917	−0.3333	−1.0583	−0.2500	−0.3833
−2.3083	−2.4417	−1.9750	−1.4250	−2.0750	−1.3750	−0.8250
0.5750	0.6542	0.4167	−0.4625	0.5000	1.3875	1.3625

Periodic trends were allowed for because the graphs for c_3, c_5 and c_6, in particular, showed distinct seasonal peaks and troughs. This was done by fitting $\cos d$ and $\sin d$ where d is day of the year starting on 29 April. Interaction terms involving years and $\cos d$ and $\sin d$ were also included. These terms allow for possible variation in the amplitude and phase in the periodic function over the three years in which the data were collected.

The regression equation was

$$\hat{c} = b_0 + b_1\xi_1 + b_2\xi_2 + b_3\xi_3 + b_4\xi_4 + b_5\cos d + b_6\sin d + b_7\,Y1 \times \cos d$$
$$+ b_8\,Y1 \times \sin d + b_9\,Y2 \times \cos d + b_{10}\,Y2 \times \sin d. \qquad (2.3)$$

The ξ_i are orthogonal polynomials replacing t, t^2, t^3 and t^4, $Y1$ is the difference between 1970/71 and 1971/72, and $Y2$ is the difference between 1972/73 and the mean of 1970/71 and 1971/72.

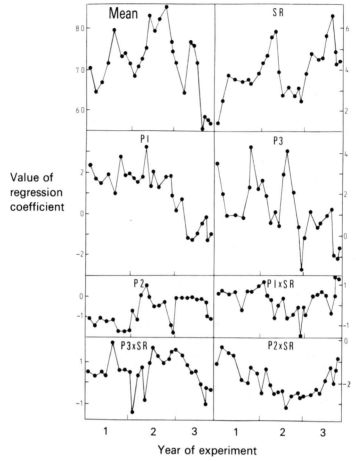

Figure 2.2 Values of the regression coefficients of Equation (2.1) for 27 weighings
for the mean and seven treatment variables (values taken from Table 2.6).

2.5 FITTING EQUATION (2.3) TO THE DATA
FOR EACH PLOT

The values of the variables $\xi_1, \ldots, Y2 \times \sin d$ on the right-hand side of
Equation (2.3) used in the analysis are listed in Table 2.7 together with the
live weights for plots 1 and 24 (Table 2.1) as examples, in the usual layout
for regression analysis.

The live weights for each plot were regressed on $\xi_1, \ldots, Y2 \times \sin d$ in
multiple regression to obtain the estimates of the coefficients $b_0, \ldots, b_{10}$ of
Equation (2.3) for each plot. These estimates are given in Table 2.8 and
become the data for the following analyses.

Table 2.7 Live weights for plots 1 and 24 and the regression variables of Equation (2.3)

		Live weights plot											
t	d	1	24	ξ_1	ξ_2	ξ_3	ξ_4	$\cos d$	$\sin d$	$Y1 \times \cos d$	$Y1 \times \sin d$	$Y2 \times \cos d$	$Y2 \times \sin d$
0	0	657	734	−0.3378	0.3954	−0.3699	0.3488	1	0	1	0	1	0
29	29	635	640	−0.3210	0.3351	−0.2403	0.1191	0.8781	0.4785	0.8781	0.4785	0.8781	0.4785
85	85	679	639	−0.2887	0.2277	−0.0396	−0.1681	0.1081	0.9941	0.1081	0.9941	0.1081	0.9941
141	141	718	696	−0.2564	0.1321	0.1017	−0.2928	−0.7547	0.6561	−0.7547	0.6561	−0.7547	0.6561
212	212	801	766	−0.2154	0.0278	0.2067	−0.2898	−0.8747	−0.4847	−0.8747	−0.4847	−0.8747	−0.4847
287	287	678	626	−0.1721	−0.0618	0.2423	−0.1736	0.2237	−0.9747	0.2237	−0.9747	0.2237	−0.9747
302	302	726	764	−0.1635	−0.0772	0.2416	−0.1434	0.4650	−0.8853	0.4650	−0.8853	0.4650	−0.8853
358	358	724	686	−0.1358	−0.1208	0.2247	−0.0422	0.9924	−0.1233	0.9924	−0.1233	0.9924	−0.1233
386	21	714	681	−0.1156	−0.1471	0.2003	0.0296	0.9366	0.3505	−0.9366	−0.3505	0.9366	0.3505
426	61	765	677	−0.0925	−0.1715	0.1622	0.1027	0.5007	0.8656	−0.5007	−0.8656	0.5007	0.8656
478	113	800	683	−0.0636	−0.1939	0.1021	0.1735	−0.3618	0.9322	0.3618	−0.9322	−0.3618	0.9322
507	142	856	722	−0.0463	−0.2023	0.0640	0.1997	−0.7638	0.6454	0.7638	−0.6454	−0.7638	0.6454
546	181	798	801	−0.0238	−0.2086	0.0103	0.2167	−0.9995	0.0306	0.9995	−0.0306	−0.9995	0.0306
581	216	810	805	−0.0030	−0.2093	−0.0397	0.2132	−0.8410	−0.5410	0.8410	0.5410	−0.8410	−0.5410
636	271	850	822	0.0287	−0.2010	−0.1128	0.1732	−0.0529	−0.9986	0.0529	0.9986	−0.0529	−0.9986
686	321	788	844	0.0587	−0.1827	−0.1726	0.1016	0.7225	−0.6914	−0.7225	0.6914	0.7225	−0.6914
728	363	808	743	0.0830	−0.1605	−0.2105	0.0260	0.9992	−0.0408	−0.9992	0.0408	0.9992	−0.0408
747	17	788	724	0.0939	−0.1483	−0.2236	−0.0114	0.9593	0.2822	0	0	−1.9187	−0.5645
776	46	722	695	0.1107	−0.1270	−0.2381	−0.0700	0.7073	0.7069	0	0	−1.4147	−1.4138
824	94	808	617	0.1384	−0.0849	−0.2448	−0.1635	−0.0401	0.9992	0	0	0.0803	−1.9984
923	193	804	736	0.1955	0.0294	−0.1733	−0.2729	−0.9851	−0.1718	0	0	1.9703	0.3437
936	206	766	734	0.2030	0.0471	−0.1538	−0.2716	−0.9225	−0.3861	0	0	1.8449	0.7721
972	242	637	695	0.2238	0.0996	−0.0858	−0.2392	−0.5270	−0.8499	0	0	1.0540	1.6998
1021	291	647	564	0.2520	0.1788	0.0426	−0.1107	0.2841	−0.9588	0	0	−0.5681	1.9176
1049	319	632	564	0.2682	0.2281	0.1362	0.0173	0.6960	−0.7181	0	0	−1.3920	1.4361
1077	347	623	551	0.2844	0.2804	0.2456	0.1931	0.9495	−0.3139	0	0	−1.8989	0.6277
1095	365		538	0.2947	0.3155	0.3247	0.3345	1	0	0	0	−2	0

Table 2.8   Estimates of the coefficients of Equation (2.3) for each plot

B	P	SR	b_0 constant	b_1 linear	b_2 quadratic	b_3 cubic	b_4 quartic	b_5 cos	b_6 sin	b_7 $Y1 \times cos$	b_8 $Y1 \times sin$	b_9 $Y1 \times cos$	b_{10} $Y2 \times sin$
1	1	1	77.6111	-0.7502	-24.5369	-12.8869	2.9218	-3.6855	-1.4583	-1.8581	0.4485	-0.2170	-0.0314
2	1	1	74.6284	3.6990	-20.9290	-13.3945	3.8809	-4.9867	-3.0619	-1.6013	0.0822	0.4016	-0.0779
3	1	1	69.4313	-15.1825	-21.5587	-13.7971	4.7755	-4.2298	-3.6018	-3.7978	-1.5405	-0.3562	2.0075
1	2	1	75.5983	-16.5442	-39.4188	-15.6451	10.4779	-3.5750	-3.4165	-1.9882	1.2695	-0.3732	1.7272
2	2	1	73.8779	-20.9678	-29.0925	-13.6589	3.9906	4.6442	-3.7798	-2.6356	-1.0778	0.0454	-1.0254
3	2	1	78.2617	-13.6616	-20.2620	-25.4956	-0.7298	-5.0596	-3.6028	-2.4801	-1.4684	-1.6446	-0.6830
1	1	2	68.1380	-4.6503	-16.0166	-13.7720	7.1146	-5.7848	-4.1829	-3.8425	1.1928	0.9032	-1.3852
2	1	2	67.8770	-15.8320	-15.0035	-18.5672	-0.8259	-2.9473	-5.9684	-1.5946	0.6672	-0.7874	-1.4961
3	1	2	64.3486	-11.8913	-20.0815	-12.0068	9.5357	-4.6008	-0.4317	-3.1099	0.4692	0.6159	-2.8675
1	2	2	66.0479	-14.3492	-24.2665	-7.4637	9.6580	-2.8215	-2.0515	-0.8676	1.2675	1.5231	-1.1783
2	2	2	69.2095	-26.8476	-18.8216	-13.2918	7.5696	-3.3505	-4.0650	-3.2946	-0.1623	-0.4055	-0.3134
3	2	2	67.2855	-20.3502	-23.5253	-14.1771	12.1175	-4.6761	-4.1145	-3.6534	0.3455	0.1299	-1.0232
1	3	1	76.2208	-18.3964	-21.8909	-11.1961	8.7494	-4.0966	-5.0609	-2.8601	0.8744	0.6183	-0.7838
2	3	1	73.3778	-9.9136	-12.3461	2.0448	14.1319	-6.3649	-2.2645	-2.1424	0.8361	2.8575	-1.4943
3	3	1	77.3481	-8.3967	-20.3459	-1.1116	11.1543	-4.0366	-2.9951	-1.1332	1.2411	1.7655	-0.5079
1	4	1	81.4763	-14.5633	-17.9630	-2.4496	5.9376	-3.1141	-4.9134	-1.2432	0.6826	1.6779	-1.6917
2	4	1	81.5380	-11.8132	-25.3019	-9.7359	9.0754	-4.1010	-4.5930	-2.3531	0.8348	1.2168	-1.5028
3	4	1	79.1666	-12.6896	-20.5162	2.4280	16.5249	-5.9969	-3.0190	-2.4750	1.4144	2.3247	-1.1815
1	3	2	69.9132	-10.6861	-27.8247	-19.1011	20.1545	-6.4174	-5.1636	-3.0048	1.5575	0.3509	-1.4488
2	3	2	72.1614	-23.4291	-23.8715	-12.5336	11.6968	-3.5308	-6.4581	-3.2924	1.1027	0.8199	-1.6437
3	3	2	70.3345	-24.6665	-26.2991	-15.9556	9.4779	-3.4877	-5.9946	-2.8194	0.4136	0.9323	-1.6337
1	4	2	68.2907	-30.3882	-20.2458	-2.4254	8.8566	-3.0190	-5.3937	-0.9836	0.7787	0.8133	-2.2410
2	4	2	66.1378	-28.1270	-15.2828	-12.3187	13.3842	-3.6599	-6.6109	-3.9893	2.2236	1.0828	-2.3084
3	4	2	70.1595	-18.0492	-21.7851	-14.9672	12.1104	-3.9701	-4.8669	-2.6393	1.7171	0.7188	-0.7035

2.6 ANALYSIS OF $b_0, \ldots, b_{10}$

The values of $b_0, \ldots, b_{10}$ were regressed in turn on the x-variables of the design matrix of Table 2.3 in univariate analysis to estimate the values of $b_0, \ldots, b_{10}$ for each treatment comparison and to test their significance. The analysis of variance was:

Source	DF
Blocks	2
P1	1
P2	1
P3	1
SR	1
P1 × SR	1
P2 × SR	1
P3 × SR	1
Error	14
Total	23

The results are given in Table 2.9. The regression coefficients on the same line as the treatment variables in the tables are interpreted as follows, using P1 as an example: 0.9372 is the mean of the 27 values of c_3 given in Table 2.6; -5.6728 is the coefficient of the linear regression of the 27 values of c_3 on time; -3.0567 is the coefficient of the quadratic regression of the 27 values of c_3 on time; and so on.

These are the coefficients that would be obtained by regressing the values of c_3 on $\xi_1, \xi_2, \ldots, Y2 \times \sin d$ directly in multiple regression. The indirect method is used because it provides the variances and covariances of the estimates as well as the estimates themselves.

The standard errors shown next to the regression coefficient were calculated as follows:

$$\operatorname{var}(b_i) = s^2 x^{ii}$$

where s^2 is the error mean square and x^{ii} is the iith element of $(X^T X)^{-1}$ of Table 2.5, then,

$$\operatorname{SE}(b_i) = [\operatorname{var}(b_i)]^{1/2}.$$

For example,

$$\operatorname{SE}(b_0) = (4.552/12)^{1/2}$$
$$= \pm 0.6159.$$

In this experiment

$$x^{ii} = 1 \Big/ \sum x_i^2$$

Table 2.9 Analysis of variance of the values of $b_0, \ldots, b_{10}$ of Table 2.6

Source	DF	b_0			b_1			b_2		
		MS	F	P(%)	MS	F	P(%)	MS	F	P(%)
Blocks	2	0.499			16.806			32.007		
P1	1	10.540	2.3	15.0	338.450	7.9	0.94	112.122	5.7	3.1
P2	1	4.579	1.0	33.3	33.461	0.79	39.0	10.989	0.56	46.6
P3	1	56.456	12.4	0.33	120.444	2.8	11.5	15.714	0.80	38.5
Stocking Rate	1	381.067	83.7	0.00	338.450	7.9	1.4	19.364	0.99	33.6
P1 × SR	1	3.952	0.87	36.7	8.014	0.19	67.2	3.872	0.20	66.3
P2 × SR	1	44.296	9.7	0.75	19.823	0.46	50.7	74.400	3.8	7.2
P3 × SR	1	3.102	0.68	42.3	35.090	0.82	38.0	128.103	6.6	2.3
Error	14	4.552			42.788			19.549		

Variable	b_0		b_1		b_2	
	Regn coefficient	Standard error	Regn coefficient	Standard error	Regn coefficient	Standard error
Constant term	72.3100	0.435	−15.3536	1.33	−21.9486	0.90
P1	0.9372	0.616	−5.6728	1.88	−3.0567	1.28
P2	−0.6178	0.616	1.6785	1.88	−0.9569	1.28
P3	1.5337	0.435	−2.2396	1.33	0.8092	0.90
Stocking Rate	3.9847	0.435	3.7553	1.33	−0.8982	0.90
P1 × SR	0.5740	0.616	−0.8172	1.88	−0.5681	1.28
P2 × SR	−1.9213	0.616	−1.2853	1.88	2.4900	1.28
P3 × SR	0.3595	0.435	1.2092	1.33	2.3103	0.90

Table 2.9 Cont.

Source	DF	b3 MS	b3 F	b3 P(%)	b4 MS	b4 F	b4 P(%)	b5 MS	b5 F	b5 P(%)
Blocks	2	3.314			5.562			0.413		
P1	1	2.437	0.08	77.6	20.492	1.1	32.1	0.370	0.25	62.7
P2	1	28.165	1.0	33.9	7.482	0.39	54.4	1.382	0.92	35.4
P3	1	246.623	8.5	1.1	208.668	10.8	0.54	0.086	0.06	81.5
Stocking Rate	1	72.738	2.5	13.5	37.399	1.9	18.6	1.318	0.88	36.5
P1 × SR	1	48.138	1.7	21.7	10.755	0.56	46.8	0.682	0.46	51.1
P2 × SR	1	25.154	0.87	36.6	1.673	0.09	77.3	0.188	0.13	72.9
P3 × SR	1	221.710	7.6	1.5	3.957	0.20	65.8	0.110	0.07	79.1
Error	14	28.837			19.365			1.501		

Variable	b3 Regn coefficient	b3 Standard error	b4 Regn coefficient	b4 Standard error	b5 Regn coefficient	b5 Standard error
Constant term	-11.3158	1.10	8.8225	0.90	-4.2565	0.250
P1	-0.4506	1.55	1.3086	1.27	0.1757	0.354
P2	-1.5320	1.55	0.7896	1.27	-0.3394	0.354
P3	3.2056	1.10	2.9486	0.90	-0.0597	0.250
Stocking rate	1.7409	1.10	-1.2483	0.90	-0.2344	0.250
P1 × SR	-2.0029	1.55	-0.9467	1.27	-0.2385	0.354
P2 × SR	1.4478	1.55	-0.3734	1.27	0.1251	0.354
P3 × SR	3.0325	1.10	0.4061	0.90	-0.0677	0.250

Table 2.9 Cont.

Source	DF	b6 MS	b6 F	b6 P(%)	b7 MS	b7 F	b7 P(%)	b8 MS	b8 F	b8 P(%)
Blocks	2	2.141			1.475			0.965		
P1	1	0.470	0.26	62.1	0.012	0.01	91.0	0.108	0.19	67.3
P2	1	0.178	0.10	76.0	0.095	0.10	75.3	0.220	0.38	54.8
P3	1	12.978	7.1	1.9	0.133	0.14	71.0	6.179	10.6	0.57
Stocking rate	1	7.577	4.1	6.2	1.773	1.9	18.8	2.647	4.5	5.1
P1 × SR	1	0.740	0.40	53.6	0.000	0.00	98.2	0.032	0.05	81.9
P2 × SR	1	0.725	0.40	54.0	0.313	0.34	57.0	0.231	0.40	53.8
P3 × SR	1	4.001	2.2	16.2	0.516	0.56	46.8	0.718	1.23	28.6
Error	14	1.834			0.927			0.582		

Variable	b6 Regn coefficient	b6 Standard error	b7 Regn coefficient	b7 Standard error	b8 Regn coefficient	b8 Standard error
Constant term	−4.0425	0.276	−2.444	0.197	0.6322	0.155
P1	−0.1979	0.391	0.0320	0.278	−0.0950	0.220
P2	0.1217	0.391	−0.0890	0.278	−0.1354	0.220
P3	−0.7354	0.276	0.0744	0.197	0.5074	0.155
Stocking rate	0.5619	0.276	0.2718	0.197	−0.3321	0.155
P1 × SR	−0.2483	0.391	−0.0065	0.278	0.0513	0.155
P2 × SR	0.2458	0.391	0.1617	0.278	0.1388	0.220
P3 × SR	0.4083	0.276	0.1467	0.197	0.1730	0.155

Table 2.9 Cont.

Source	DF	b_9 MS	b_9 F	b_9 P(%)	b_{10} MS	b_{10} F	b_{10} P(%)
Blocks	2	0.025			0.481		
P1	1	0.137	0.22	64.8	0.061	0.09	77.1
P2	1	0.020	0.03	86.1	0.373	0.54	47.5
P3	1	9.809	15.6	0.15	5.315	7.7	1.5
Stocking rate	1	0.109	0.17	68.4	6.508	9.4	0.84
P1 × SR	1	0.447	0.71	41.4	1.774	2.6	13.2
P2 × SR	1	0.023	0.04	84.9	0.094	0.14	71.8
P3 × SR	1	4.055	6.4	2.4	1.963	2.8	11.5
Error	14	0.631			0.693		

Variable	b_9 Regn coefficient	b_9 Standard error	b_{10} Regn coefficient	b_{10} Standard error
Constant term	0.6256	0.162	−0.9578	0.170
P1	−0.1071	0.229	0.0712	0.240
P2	−0.0408	0.229	0.1764	0.240
P3	0.6393	0.162	−0.4706	0.170
Stocking rate	0.0675	0.162	0.5207	0.170
P1 × SR	−0.1931	0.229	−0.3844	0.240
P2 × SR	0.0445	0.229	0.0886	0.240
P3 × SR	0.4111	0.162	−0.2860	0.170

Table 2.10 Error sums of squares and products matrix for the multivariate analysis of $b_0, b_1, \ldots, b_{10}$ of Table 2.5

b_0	b_1	b_2	b_3	b_4	b_5	b_6	b_7	b_8	b_9	b_{10}
63.722										
62.067	59.904									
14.656	−0.72710	273.69								
−68.676	−34.591	−7.3028	403.72							
−67.316	139.18	−103.23	103.89	271.11						
8.0845	−58.373	−17.419	−4.5102	−49.237	21.015					
−11.398	55.315	−12.075	5.2290	41.769	−7.1804	25.672				
8.7309	20.813	−2.6331	21.099	−23.632	6.9229	3.5019	12.982			
0.73660	37.961	−22.560	5.1419	28.056	−1.7747	4.1329	0.29109	8.1537		
−11.482	24.013	−3.1302	45.967	24.429	−4.0975	9.0867	1.2837	3.0873	8.8316	
1.2984	−14.448	−21.177	−9.9823	10.843	2.3024	−5.5806	−2.4282	2.3064	−1.9507	9.7024

because the design is orthogonal. The x_i are the values in the ith column of Table 2.3.

The coefficients $b_0, \ldots, b_{10}$ were also regressed on the x-variable of Table 2.3 in multivariate analysis to obtain the error sums of squares and products matrix in Table 2.10, and from this S_b, the estimate of the variance–covariance matrix of $b_0, \ldots, b_{10}$.

2.7 CHOOSING EQUATIONS TO APPROXIMATE TRENDS IN $c_3 \ldots, c_9$

All the trend coefficients including the mean are given in Table 2.11. Asterisks have been used to indicate individual coefficients that were significant at $P = 0.05$ or nearly so.

P1

None of the periodic coefficients was significant but the linear and quadratic were. A quadratic equation is likely to be sufficient but this needs to be checked by regressing the live weights on ξ_1, ξ_2, ξ_3 and ξ_4 only and analysing the coefficients obtained. This step, which is carried out in section 2.7.1, is necessary because the polynomials and the periodic terms are non-orthogonal and the coefficients obtained depend on which other variables are included in the regression.

At this stage of the analysis there did not appear to be any significant trend in time in the values of $P1 \times SR$.

P2

There did not appear to be any important trend in P2 as no coefficient was significant. However, the effect of P2 should be examined at each stocking rate because the mean was significant for $P2 \times SR$ and the quadratic nearly so, $P = 0.072$. This is done in section 2.7.2.

P3

Almost all terms appeared to be required to describe the trends in P3 over time. Although b_1 and b_2 were not significant, the linear and quadratic terms would be included because the higher-order coefficients b_3 and b_4 were significant; the $Y1 \times \cos d$ term could be omitted. However, the significant coefficients for $P3 \times SR$ indicate that P3 should also be examined at each stocking rate. The final decision on the equation will be left until this is done in section 2.7.2.

Table 2.11   Estimates of $b_0, b_1, ..., b_{10}$ of Equation (2.3) for the constant term and

Factor				Regression coefficients	
	b_0	$b_1(\xi_1)$	$b_2(\xi_2)$	$b_3(\xi_3)$	$b_4(\xi_4)$
Constant term	72.3100*	−15.3536*	−21.9486*	−11.3158*	8.8225*
P1	0.9372	−5.6728*	−3.0567*	−0.4506	1.3086
P2	−0.6178	1.6785	−0.9569	−1.5320	0.7896
P3	1.5337*	−2.2396	0.8092	3.2056*	2.9486*
Stocking rate	3.9847*	3.7553*	−0.8982	1.7409	−1.2483
P1 × SR	0.5740	−0.8172	−0.5681	−2.0029	−0.9467
P2 × SR	−1.9213*	−1.2853	2.4900	1.4478	−0.3734
P3 × SR	0.3595	1.2092	2.3103	3.0325*	0.4061

*$P < 0.05$.

SR

The effect of stocking rate appeared to increase linearly over the period. Although $Y1 \times \sin d$ and $Y2 \times \sin d$ were the only significant periodic terms, $\sin d$ must be retained to obtain the correct coefficients for each year; $\cos d$ may be important when $Y1 \times \cos d$ and $Y2 \times \cos d$ are omitted.

Therefore, the following regression equation was fitted to the values of the stocking rate comparison:

$$\hat{c}_6 = b_0 + b_1\xi_1 + b_5\cos d + b_6 \sin d + b_8\ Y1 \times \sin d + b_{10}\ Y2 \times \sin d. \quad (2.4)$$

This equation was fitted in two stages as usual.

Stage 1

The live weights for each plot were regressed on the values of ξ_1, $\cos d$, $\sin d$, $Y1 \times \sin d$ and $Y2 \times \sin d$ in Table 2.7. The coefficients obtained are given in Table 2.12.

Stage 2

The regression coefficients in Table 2.12 were regressed on the x-variable of Table 2.3. This procedure fits the regression Equation (2.4) to the values of every comparison in the analysis of variance. However, the equation obtained for the stocking rate comparison is the only one to be considered. The analyses are given in Table 2.13.

The constant, b_0, $b_8(Y1 \times \sin d)$ and $b_{10}(Y2 \times \sin d)$ were very highly significant, indicating that the periodic trends differed from year to year. Although b_6 was not significant, the $\sin d$ term must be retained to provide the correct coefficients for each year in conjunction with b_8 and b_{10}. The

each treatment comparison of Equation (2.1)

with variables in brackets					
$b_5(\cos d)$	$b_6(\sin d)$	$b_7(Y1 \times \cos d)$	$b_8(Y1 \times \sin d)$	$b_9(Y2 \times \cos d)$	$b_{10}(Y2 \times \sin d)$
-4.2565^*	-4.0425^*	-2.4415^*	0.6323^*	0.6256^*	-0.9578^*
0.1757	-0.1979	0.0321	-0.0950	-0.1071	0.0712
-0.3394	0.1217	-0.0890	-0.1354	-0.0408	0.1764
-0.0597	-0.7354^*	0.0744	0.5074^*	0.6393^*	-0.4706^*
-0.2344	0.5619	0.2718	-0.3321^*	0.0675	0.5207^*
-0.2385	-0.2483	-0.0065	0.0513	-0.1931	-0.3844
0.1251	0.2458	0.1617	0.1388	0.0445	0.0886
-0.0677	0.4083	0.1467	0.1730	0.4111^*	-0.2860

Table 2.12   Estimates of the coefficients of Equation (2.4) for each plot

B P SR	b_0 constant	b_1 linear	b_5 cos	b_6 sin	b_8 $Y1 \times \sin$	b_{10} $Y2 \times \sin$
1 1 1	75.1888	4.6098	-5.4082	0.2541	0.4825	-3.0486
2 1 1	75.0864	7.4157	-6.4234	-1.5943	0.5307	-2.6743
3 1 1	69.7948	-8.5493	-5.4028	-1.6898	-0.9636	-0.7645
1 2 1	76.4491	-8.7703	-5.7895	-0.8642	0.5115	-2.8913
2 2 1	74.4707	-15.0214	-6.6225	-1.9029	-1.2903	-4.4846
3 2 1	79.0667	-4.8008	-6.7398	-0.8367	1.0967	-4.0482
1 1 2	68.2607	-1.1981	-6.4660	-2.7224	2.2756	-3.4301
2 1 2	68.4546	-10.2486	-4.2877	-4.0966	2.4816	-3.9229
3 1 2	64.5547	-7.7248	-5.3385	1.1222	0.9067	-4.7717
1 2 2	66.3676	-13.1459	-4.0813	-1.0942	0.4396	-3.6738
2 2 2	69.5005	-20.7530	-4.0186	-2.1960	0.6670	-2.7523
3 2 2	67.5673	-14.2799	-5.3813	-2.0725	0.9003	-3.9264
1 3 1	76.5227	-14.5528	-5.0945	-3.5815	0.9630	-3.3346
2 3 1	73.0108	-12.3457	-6.1855	-2.4166	-0.3965	-2.1860
3 3 1	77.4211	-8.3770	-4.7656	-2.5505	-0.2693	-2.2937
1 4 1	81.5870	-14.5984	-4.1468	-4.5674	-0.4520	-3.3229
2 4 1	81.8361	-8.8254	-5.3951	-3.2948	0.3551	-4.2399
3 4 1	78.9633	-12.9961	-6.1597	-2.7230	-0.5826	-2.7318
1 3 2	70.2878	-4.3648	-6.8385	-2.5757	2.6367	-4.9556
2 3 2	72.3982	-18.9763	-4.3877	-4.7616	1.3149	-4.4125
3 3 2	70.7343	-20.2382	-4.8409	-4.1526	0.9233	-4.7856
1 4 2	68.5084	-28.6523	-3.8735	-4.6477	-0.5260	-4.2143
2 4 2	66.0816	-24.6388	-3.6449	-5.1403	3.2466	-4.1772
3 4 2	70.4413	-13.8409	-4.6823	-3.0703	2.5148	-3.3921

linear coefficient, b_1, and b_5 (cos d) were also significant. The fitted equation was:

$$\hat{c}_6 = 4.010 + 3.385\,\xi_1 - 0.429 \cos d + 0.402 \sin d - 0.742\ Y1 \times \sin d$$
$$+ 0.516\ Y2 \times \sin d. \tag{2.5}$$

Equation (2.5) will be used in section 2.8 to provide expected values for c_6 at each weighing time.

2.7.1 Fitting polynomial trends only

The live weights for each plot were regressed on the values of ξ_1, ξ_2, ξ_3 and ξ_4 of Table 2.7. The coefficients obtained for each plot are listed in Table 2.14. The values of each coefficient were regressed on the values of the x-variables of Table 2.3 in univariate and multivariate analyses. These analyses are given in Tables 2.15 and 2.16. The P1 component and its interaction with SR are to be considered.

P1

The effect of P1 was taken to be independent of stocking rate, because no coefficient was significant for P1 $\times$ SR. A quadratic was required to approximate the trends in the c_3 because b_1 and b_2 were both significant for P1. The fitted regression was:

$$\hat{c}_3 = \ 0.996 \ -5.285\xi_1 - 2.716\xi_2.$$
$$(\pm 0.65)\ \ (\pm 1.69)\ \ (\pm 1.15) \tag{2.6}$$

2.7.2 Values of P2 and P3 at each stocking rate

The values of P2 and P3 at each stocking rate can be obtained conveniently in a similar manner to those of P2, P3 and P2 $\times$ SR and P3 $\times$ SR obtained in section 2.2. The regression equation to be fitted is now

$$\hat{y} = c_0 x_0 + c_1 x_1 + c_2 x_2 + c_3 x_3 + c_7 x_4 + c_{4.1} x_5 + c_{4.2} x_6 + c_{5.1} x_7$$
$$+ c_{5.2} x_8 + c_6 x_9 \tag{2.7}$$

instead of (2.1). The subscripts used for the regression coefficients in Equation (2.1) have been retained in (2.7) where possible. The values of the x-variables used to fit Equation (2.7) are given in Table 2.17.

Table 2.13 Analysis of variance of the values of b_0, b_1, b_5, b_6, b_8, b_{10} of Table 2.12

Source	DF	MS	b_0 F	P(%)	MS	b_1 F	P(%)	MS	b_5 F	P(%)
Blocks	2	0.672			16.195			0.1800		
P1	1	12.144			310.856			0.0401		
P2	1	4.133			50.829			1.4773		
P3	1	45.431			337.051			1.4724		
Stocking rate	1	385.844	76.1	0.00	275.064	7.0	2.0	4.4137	5.3	3.7
P1 × SR	1	5.020			0.781			1.7089		
P2 × SR	1	47.286			41.844			1.0339		
P3 × SR	1	1.283			0.432			0.4630		
Error	14	5.068			39.551			0.8278		
Variable		Regn coefficient	Standard error		Regn coefficient	Standard error		Regn coefficient	Standard error	
Stocking rate		4.0096	0.460		3.3854	1.28		−0.4288	0.186	

Source	DF	MS	b_6 F	P(%)	MS	b_8 F	P(%)	MS	b_{10} F	P(%)
Blocks	2	2.812			0.193			0.1920		
P1	1	0.005			0.957			0.8345		
P2	1	0.966			0.032			0.0010		
P3	1	27.711			0.119			0.5574		
Stocking rate	1	3.872	2.7	12.5	13.199	10.9	0.53	6.4006	10.9	0.52
P1 × SR	1	0.069			1.284			3.7508		
P2 × SR	1	0.037			0.149			1.9605		
P3 × SR	1	0.026			0.424			0.4436		
Error	14	1.454			1.213			0.5860		
Variable		Regn coefficient	Standard error		Regn coefficient	Standard error		Regn coefficient	Standard error	
Stocking rate		0.4017	0.246		−0.7416	0.225		0.5164	0.156	

Table 2.14 Estimates of the coefficients for the regression of live weight on ξ_1, ξ_2, ξ_3 and ξ_4 of Table 2.7 for each plot

B P SR	b_0 constant	b_1 linear	b_2 quadratic	b_3 cubic	b_4 quartic
1 1 1	73.9569	2.0827	−28.0399	−11.8827	−4.5615
2 1 1	73.8163	7.0618	−25.8115	−12.3714	−6.8627
3 1 1	68.9561	−7.5692	−22.8830	−6.0796	−6.8217
1 2 1	75.2235	−10.0245	−38.2467	−12.4141	2.4101
2 2 1	73.0379	−16.5421	−36.4019	−10.6608	−7.0662
3 2 1	77.5389	−6.8782	−26.9339	−18.6740	−9.0743
1 1 2	66.9017	−0.7347	−24.0691	−15.2808	−7.8301
2 1 2	67.4312	−8.6322	−19.4417	−15.6381	−7.1879
3 1 2	63.0867	−12.3425	−30.1116	−15.2107	−1.5639
1 2 2	65.3308	−14.6869	−28.8818	−11.8436	1.4113
2 2 2	68.6266	−20.4612	−22.3101	−9.5126	−1.7362
3 2 2	66.3308	−14.9647	−30.1795	−12.6166	−0.3372
1 3 1	76.4639	−13.2281	−26.6985	−10.8128	−2.5506
2 3 1	71.9003	−11.5798	−22.1287	−4.3238	−3.1966
3 3 1	76.5264	−7.5008	−24.7487	−4.6825	−0.0128
1 4 1	80.7707	−12.3614	−23.8488	−5.1126	−4.4264
2 4 1	80.6345	−8.5177	−31.9111	−11.4298	−3.0632
3 4 1	77.8150	−12.3174	−28.6734	−2.7806	0.0549
1 3 2	68.6689	−5.1153	−35.8446	−19.6548	5.4436
2 3 2	71.3715	−17.1384	−29.7906	−12.5193	−0.6230
3 3 2	69.5792	−19.3562	−32.6120	−15.6904	−2.3332
1 4 2	67.6408	−26.9676	−26.6028	−3.9403	0.4801
2 4 2	65.1531	−22.0179	−22.2005	−14.5865	−0.2774
3 4 2	69.3682	−12.9193	−25.9048	−15.7764	0.6057

The analysis of variance used to test the significance of the treatment effects was:

Source	DF
Blocks	2
P1	1
P1 × SR	1
P2/SR1	1
P2/SR2	1
P3/SR1	1
P3/SR2	1
SR	1
Error	14
Total	23

(cf. section 2.6). P2/SR1 stands for P2 at stocking rate 1.

The only difference between this analysis and that used previously is the different partitioning of the sources of variation associated with P2 and P3 and their interactions with stocking rate. Equation (2.7) was fitted to the data at each weighing. The estimates for P2/SR1, P2/SR2, P3/SR1 and P3/SR2 are given in Table 2.18 and are plotted in Figure 2.3.

2.7.3 Trends in the values of P2 and P3 at each stocking rate

The values of $b_0, \ldots, b_{10}$ of Table 2.8 were regressed on the x-variables of the new design matrix of Table 2.17 to estimate trends in P2/SR1, P2/SR2, P3/SR1 and P3/SR2 and to test their significance. The analyses are given in Table 2.19 and are summarized in Table 2.20.

The periodic terms did not appear to be needed for P2/SR1 or P2/SR2. The values of b_0, b_1, b_2, b_3 and b_4 of Table 2.14 were regressed on the x-variables of Table 2.17 to estimate polynomial trends in P2/SR1 and P2/SR2. These analyses are given in Table 2.21.

P2/SR1

The mean of the values of $c_{4.1}$ was significantly different from 0, $P=0.015$ (Table 2.21), but none of the other coefficients was significant. The regression was

$$\hat{c}_{4.1} = -2.55(\pm 0.92). \qquad (2.8)$$

P2/SR2

The quadratic coefficient, b_2, was significant, $P=0.030$. The regression was

$$\hat{c}_{4.2} = \begin{array}{ccc} 1.243 & +3.382\xi_1 & -3.923\xi_2. \\ (\pm 0.92) & (\pm 2.4) & (\pm 1.6) \end{array} \qquad (2.9)$$

P3/SR1

All polynomial coefficients except b_1 were significant (Table 2.20) and all polynomial terms were retained. The interaction coefficients b_7, b_8 and b_9 were significant and were retained together with $\cos d$ and $\sin d$ to provide the correct coefficients for $\cos d$ and $\sin d$ in the three years. $Y1 \times \cos d$ was also retained although it could have been dropped; refitting to exclude one term out of eleven did not seem worth while. The regression was

$$
\begin{aligned}
\hat{c}_{5.1} = \quad & 1.893 \; -1.030\xi_1 +3.120\xi_2 +6.238\xi_3 +3.355\xi_4 \\
& (\pm 0.62) \quad (\pm 1.9) \quad (\pm 1.3) \quad (\pm 1.6) \quad (\pm 1.3) \\[4pt]
& - \; 0.127 \;\; \cos d - \; 0.327 \;\; \sin d + \; 0.221 \;\; Y1 \times \cos d \\
& (\pm 0.36) \qquad\quad (\pm 0.39) \qquad\quad (\pm 0.28) \\[4pt]
& + \; 0.680 \;\; Y1 \times \sin d + \; 1.050 \;\; Y2 \times \cos d - \; 0.757 \;\; Y2 \times \sin d. \\
& (\pm 0.22) \qquad\qquad (\pm 0.23) \qquad\qquad (\pm 0.24)
\end{aligned}
\qquad (2.10)
$$

Table 2.15   Analysis of variance of the values of b_0, b_1, b_2, b_3 and b_4 of Table 2.14

Source	DF	b_0		
		MS	F	P(%)
Blocks	2	0.715		
P1	1	11.894	2.3	15.0
P2	1	5.164	1.0	33.3
P3	1	50.020		
Stocking rate	1	385.150		
P1 × SR	1	3.198	0.62	44.3
P2 × SR	1	43.271	8.4	1.2
P3 × SR	1	1.761		
Error	14	5.137		
Variable		Regn coefficient		Standard error
P1		0.9956		0.654
P1 × SR		0.5163		0.654

Source	DF	b_3		
		MS	F	P(%)
Blocks	2	0.012		
P1	1	0.046	0.00	96.0
P2	1	16.467	0.95	34.6
P3	1	39.720		
Stocking rate	1	108.568		
P1 × SR	1	46.303	2.7	12.4
P2 × SR	1	14.225	0.82	38.0
P3 × SR	1	51.059		
Error	14	17.319		
Variable		Regn coefficient		Standard error
P1		0.0618		1.20
P1 × SR		−1.9643		1.20

P3/SR2

The sine coefficient, b_6, was highly significant, $P = 0.011$, but none of the polynomial coefficients was significant although the P-values for b_0, b_1 and b_4 were low. A simple periodic curve involving the mean, $\cos d$ and $\sin d$ may be adequate. However, the within-plot regressions should be re-fitted using only periodic terms to decide this. This is done next.

2.7.4 Fitting periodic trends only

The live weights for each plot were regressed on the values of $\cos d$, $\sin d$, $Y1 \times \cos d$, $Y1 \times \sin d$, $Y2 \times \cos d$ and $Y2 \times \sin d$ given in Table 2.7. The

MS	b_1 F	P(%)	MS	b_2 F	P(%)
9.624			15.488		
335.212	9.8	0.7	88.548	5.6	3.3
37.392	1.1	31.3	13.402	0.85	37.3
117.824			0.229		
253.254			2.924		
3.648	0.11	74.8	24.365	1.5	23.5
31.386	0.92	35.4	98.593	6.2	2.6
0.157			61.022		
34.115			15.852		
Regn coefficient		Standard error	Regn coefficient		Standard error
−5.2853		1.69	−2.7164		1.15
−0.5514		1.69	−1.4249		1.15

MS	b_4 F	P(%)
12.997		
34.800	3.7	7.5
0.937	0.10	75.7
64.424		
40.619		
10.838	1.2	30.1
0.000	0.00	99.9
0.129		
9.396		
Regn coefficient		Standard error
1.7029		0.885
−0.9504		0.885

Table 2.16 Error sums of squares and products matrix for the multivariate analysis of the value of b_0, b_1, b_2, b_3 and b_4 of Table 2.12

b_0	b_1	b_2	b_3	b_4
71.907				
78.360	477.56			
40.467	−17.478	221.66		
−40.474	−223.68	20.924	24.300	
−30.418	29.804	−104.84	9.4745	131.85

Table 2.17   Design matrix used to fit Equation (2.7)

Block	P	SR	x_0	B1 x_1	B2 x_2	P1 x_3	P1 × SR x_4
1	S	8.9	1	1		1	1
	S	13.3	1	1		1	−1
	SP	8.9	1	1		−1	−1
	SP	13.3	1	1		−1	1
	SPL	8.9	1	1			
	SPL	13.3	1	1			
	SL	8.9	1	1			
	SL	13.3	1	1			
2	S	8.9	1		1	1	1
	S	13.3	1		1	1	−1
	SP	8.9	1		1	−1	−1
	SP	13.3	1		1	−1	1
	SPL	8.9	1		1		
	SPL	13.3	1		1		
	SL	8.9	1		1		
	SL	13.3	1		1		
3	S	8.9	1	−1	−1	1	1
	S	13.3	1	−1	−1	1	−1
	SP	8.9	1	−1	−1	−1	−1
	SP	13.3	1	−1	−1	−1	1
	SPL	8.9	1	−1	−1		
	SPL	13.3	1	−1	−1		
	SL	8.9	1	−1	−1		
	SL	13.3	1	−1	−1		
Sum of squares			24	16	16	12	12

Table 2.18   Estimates of $c_{4.1}$, $c_{4.2}$, $c_{5.1}$, $c_{5.2}$ of Equation (2.7) for 27 weighing times

Variable and coefficient	1	2	3	4	5	6
P2/SR1 $c_{4.1}$	−2.2500	−1.6667	−1.7333	−1.9833	−2.9667	−3.6667
P2/SR2 $c_{4.2}$	−0.0333	−1.2000	−0.5500	−0.6000	0.7000	0.4000
P3/SR1 $c_{5.1}$	3.8417	2.7167	1.4833	1.8250	0.5667	3.0333
P3/SR2 $c_{5.2}$	3.0250	1.3833	0.4833	0.5000	1.0417	1.6500

Table 2.18   cont.

Variable and coefficient	15	16	17	18	19	20
P2/SR1 $c_{4.1}$	−3.6000	−2.9333	−2.8500	−3.5167	−2.7167	−2.7167
P2/SR2 $c_{4.2}$	2.7833	2.4000	2.0333	1.9833	2.5833	2.4667
P3/SR1 $c_{5.1}$	3.4583	1.4500	0.2333	−3.1417	−0.2583	1.5917
P3/SR2 $c_{5.2}$	4.9083	2.5583	0.4833	−0.2750	−0.1000	0.7750

P2/SR1 x_5	P2/SR2 x_6	P3/SR1 x_7	P3/SR2 x_8	SR x_9
		−1		1
			−1	−1
		−1		1
			−1	−1
1		1		1
	1		1	−1
−1		1		1
	−1		1	−1
		−1		1
			−1	−1
		−1		1
			−1	−1
1		1		1
	1		1	−1
−1		1		1
	−1		1	−1
		−1		1
			−1	−1
		−1		1
			−1	−1
1		1		1
	1		1	−1
−1		1		1
	−1		1	−1
6	6	12	12	24

Weighing time 7	8	9	10	11	12	13	14
−2.9333	−3.2000	−3.2000	−2.4833	−2.0167	−1.9500	−2.4500	−2.9500
−0.5000	−0.0500	1.8667	0.1000	2.1333	3.1333	2.4500	1.8000
4.9833	3.0667	3.6167	2.4250	0.7167	0.4167	0.2083	3.3750
3.7417	1.3167	1.7583	1.4917	0.2917	1.9583	0.2167	2.6000

Weighing time 21	22	23	24	25	26	27
−2.4333	−2.4500	−2.1667	−1.6167	−2.3333	−2.3167	−1.8833
2.1833	2.4333	1.7833	1.2333	1.8167	0.4333	−0.2333
0.9417	1.2333	1.3750	0.7833	1.4833	2.1917	2.7000
−0.2083	−0.0750	0.5417	1.7083	0.4833	−0.5833	−0.0250

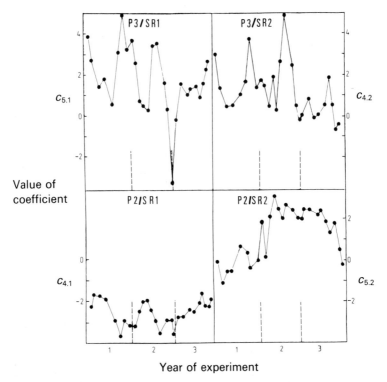

Figure 2.3 Values of $c_{4.1}$, $c_{4.2}$, $c_{5.1}$ and $c_{5.2}$ of Equation (2.6) for 27 weighing times (values taken from Table 2.18).

regression coefficients obtained for each plot were then regressed on the x-variables of Table 2.17.

P3/SR2

$Y1 \times \sin d$ and $Y2 \times \sin d$ were omitted from the regression because b_8 and b_{10} were non-significant for P3/SR2. The live weights for each plot were regressed on the values of $\cos d$, $\sin d$, $Y1 \times \cos d$ and $Y2 \times \cos d$ of Table 2.7. The estimates of the coefficients for this regression for each plot are given in Table 2.22. These estimates were regressed on the x-variables of Table 2.17. These analyses are given in Tables 2.23 and 2.24; all terms were retained because b_6 and b_7 were significant and b_9 was nearly so. The equation was

$$\hat{c}_{5.2} = \underset{(\pm 0.64)}{1.10} + \underset{(\pm 0.27)}{0.21} \cos d - \underset{(\pm 0.32)}{0.74} \sin d + \underset{(\pm 0.18)}{0.43} Y1 \times \cos d$$

$$+ \underset{(\pm 0.14)}{0.27} Y2 \times \cos d.$$

$$(2.11)$$

2.8 PREDICTED VALUES OF TREATMENT COMPARISONS

Predicted values of the comparisons are calculated from the regressions on time. The values of the appropriate variables of Table 2.7 are substituted in Equations (2.5), (2.6), (2.8), (2.9), (2.10) and (2.11). For example, the predicted values of $c_{5.2}$ are given by Equation (2.11) as:

$$\hat{c}_{5.2} = 1.097 + 0.208 \cos d - 0.738 \sin d + 0.430 \, Y1 \times \cos d + 0.270 \, Y2 \times \cos d.$$

The predicted value for the third weighing, for example, is then

$$\hat{c}_{5.2} = 1.097 + 0.208 \times 0.1081 - 0.738 \times 0.9941 + 0.430 \times 0.1081 + 0.270 \times 0.1081$$
$$= 0.46.$$

The variance of $\hat{c}_{5.2}$ is given by expression (2.12) writing t_5, t_6, t_7, t_9 for $\cos d$, $\sin d$, $Y1 \times \cos d$ and $Y2 \times \cos d$.

$$\begin{aligned}
\operatorname{var}(\hat{c}_{5.2}) = {} &\operatorname{var}(b_0) + t_5^2 \operatorname{var}(b_5) + t_6^2 \operatorname{var}(b_6) + t_7^2 \operatorname{var}(b_7) + t_9^2 \operatorname{var}(b_9) \\
&+ 2t_5 \operatorname{cov}(b_0 b_5) + 2t_6 \operatorname{cov}(b_0 b_6) + 2t_7 \operatorname{cov}(b_0 b_7) + 2t_9 \operatorname{cov}(b_0 b_9) \\
&+ 2t_5 t_6 \operatorname{cov}(b_5 b_6) + 2t_5 t_7 \operatorname{cov}(b_5 b_7) + 2t_5 t_9 \operatorname{cov}(b_5 b_9) \\
&+ 2t_6 t_7 \operatorname{cov}(b_6 b_7) + 2t_6 t_9 \operatorname{cov}(b_6 b_9) + 2t_7 t_9 \operatorname{cov}(b_7 b_9). \quad (2.12)
\end{aligned}$$

This follows from the general form of Equation (2.11), namely:

$$\hat{c}_{5.2} = b_0 + b_5 \cos d + b_6 \sin d + b_7 \, Y1 \times \cos d + b_9 \, Y2 \times \cos d.$$

Expression (2.12) is evaluated by substituting in it the values of $\cos d$, $\sin d$, $Y1 \times \cos d$ and $Y2 \times \cos d$ from Table 2.7 and the values of the variances and covariances of b_0, b_5, b_6, b_7 and b_9. The variances and covariances are the elements of S_b obtained from the matrix of sums of squares and products in Table 2.24 by dividing by 14, the number of degrees of freedom for error.

The expressions for the predicted values and their variances, which can be very long, are simplified in matrix notation.

Define the vector t by

$$t^{\mathrm{T}} = [1, \cos d, \sin d, Y1 \times \cos d, Y2 \times \cos d]$$

and b by

$$b^{\mathrm{T}} = [b_0, b_5, b_6, b_7, b_9].$$

Then

$$\hat{c}_{5.2} = [1, \cos d, \sin d, Y1 \times \cos d, Y2 \times \cos d] \begin{bmatrix} b_0 \\ b_5 \\ b_6 \\ b_7 \\ b_9 \end{bmatrix}$$

$$= t^{\mathrm{T}} b \qquad (2.13)$$

Table 2.19   Analysis of variance of the values of $b_0, b_1, \ldots, b_{10}$ of Table 2.8

Source	DF	MS	b_0 F	P(%)
Blocks	2	0.500		
Stocking rate	1	381.067		
P1	1	10.540		
P1 × SR	1	3.953		
P2/SR1	1	38.680	8.5	1.1
P2/SR2	1	10.195	2.2	15.7
P3/SR1	1	43.012	9.5	0.83
P3/SR2	1	16.545	3.6	7.7
Error	14	4.552		

Variable	Regn coefficient	Standard error
P2/SR1	−2.5390	0.871
P2/SR2	1.3035	0.871
P3/SR1	1.8932	0.616
P3/SR2	1.1742	0.616

Source	DF	MS	b_3 F	P(%)
Blocks	2	3.314		
Stocking rate	1	72.738		
P1	1	2.437		
P1 × SR	1	48.138		
P2/SR1	1	0.043	0.00	97.0
P2/SR2	1	53.276	1.8	19.6
P3/SR1	1	466.973	16.2	0.13
P3/SR2	1	0.359	0.01	91.3
Error	14	28.837		

Variable	Regn coefficient	Standard error
P2/SR1	−0.0842	2.19
P2/SR2	−2.9798	2.19
P3/SR1	6.2381	1.55
P3/SR2	0.1731	1.55

MS	b_1 F	P%	MS	b_2 F	P(%)
16.806			32.007		
338.450			19.364		
386.164			112.122		
8.014			3.872		
0.928	0.02	88.5	14.101	0.72	41.0
52.704	1.2	28.6	71.288	3.6	7.7
12.742	0.30	59.4	116.775	6.0	2.8
142.730	3.3	8.9	27.042	1.4	25.9
42.788			19.549		
Regn coefficient	Standard error		Regn coefficient	Standard error	
0.3932	2.67		1.5330	1.81	
2.9638	2.67		-3.4469	1.81	
-1.0305	1.89		3.1195	1.28	
-3.4488	1.89		-1.5012	1.28	

MS	b_4 F	P%	MS	b_5 F	P(%)
5.562			0.413		
37.399			1.318		
20.492			0.370		
10.755			0.682		
1.040	0.05	82.0	0.276	0.18	67.5
8.115	0.42	52.8	1.294	0.86	36.9
135.049	7.0	1.9	0.195	0.13	72.4
77.576	4.0	6.5	0.001	0.00	98.2
19.365			1.501		
Regn coefficient	Standard error		Regn coefficient	Standard error	
0.4163	1.80		-0.2143	0.500	
1.1630	1.80		-0.4645	0.500	
3.3547	1.27		-0.1274	0.354	
2.5426	1.27		0.0080	0.354	

Table 2.19   (cont).

Source	DF	MS	b_6 F	P(%)
Blocks	2	2.141		
Stocking rate	1	7.577		
P1	1	0.470		
P1 × SR	1	0.740		
P2/SR1	1	0.810	0.44	51.7
P2/SR2	1	0.092	0.05	82.6
P3/SR1	1	1.284	0.70	41.7
P3/SR2	1	15.695	8.6	1.1
Error	14	1.834		

Variable	Regn coefficient	Standard error
P2/SR1	0.3675	0.553
P2/SR2	−0.1241	0.553
P3/SR1	−0.3271	0.391
P3/SR2	−1.1437	0.391

Source	DF	MS	b_9 F	P(%)
Blocks	2	0.0253		
Stocking rate	1	0.1093		
P1	1	0.1376		
P1 × SR	1	0.4472		
P2/SR1	1	0.0001	0.00	99.1
P2/SR2	1	0.0437	0.07	79.6
P3/SR1	1	13.2399	21.0	0.04
P3/SR2	1	0.6251	0.99	33.6
Error	14	0.6308		

Variable	Regn coefficient	Standard error
P2/SR1	0.0037	0.324
P2/SR2	−0.0853	0.324
P3/SR1	1.0504	0.229
P3/SR2	0.2282	0.229

	b_7				b_8	
MS	F	P%	MS	F	P(%)
1.4750			0.9651		
1.7731			2.6475		
0.0123			0.1084		
0.0005			0.0317		
0.0310	0.03	85.6	0.0000	0.00	99.2
0.3772	0.41	53.4	0.4513	0.78	39.4
0.5870	0.63	44.0	5.5555	9.5	0.80
0.0625	0.07	79.9	1.3422	2.3	15.1
0.9273			0.5824		

Regn coefficient	Standard error	Regn coefficient	Standard error
0.0726	0.393	0.0033	0.312
−0.2507	0.393	−0.2743	0.312
0.2212	0.278	0.6804	0.220
−0.0722	0.278	0.3344	0.220

	b_{10}	
MS	F	P(%)
0.4811		
6.5082		
0.0608		
1.7736		
0.4214	0.61	44.9
0.0462	0.07	80.0
6.8692	9.9	0.71
0.4090	0.59	45.5
0.6930		

Regn coefficient	Standard error
0.2650	0.340
0.0878	0.340
−0.7566	0.240
−0.1846	0.240

Table 2.20 Estimates of $b_0, b_1, \ldots, b_{10}$ of Equation (2.3) for P2/SR1, P2/SR2, P3/SR1 and P3/SR2

		b_0	b_1	b_2	b_3	b_4	b_5	b_6	b_7	b_8	b_9	b_{10}
P2/SR1	$c_{4.1}$	−2.5390*	0.3932	1.5330	−0.0842	0.4163	−0.2143	0.3675	0.0726	0.0033	0.0037	0.2650
P2/SR2	$c_{4.2}$	1.3035	2.9638	−3.4469	−2.9798	1.1630	−0.4645	−0.1241	−0.2507	−0.2450	−0.0907	0.0853
P3/SR1	$c_{5.1}$	1.8932*	−1.0305	3.1195*	6.2381*	3.3547*	−0.1274	−0.3271	0.2212	0.6804*	1.0504*	−0.7566*
P3/SR2	$c_{5.2}$	1.1742	−3.4488	−1.5012	0.1731	2.5426	0.0080	−1.1437*	−0.0722	0.3344	0.2282	−0.1846

*$P = <0.05$.

and

$$\text{var}(c) = t^T V(b) t \tag{2.14}$$

where

$$V(b) = \begin{bmatrix} \text{var}(b_o) & \text{cov}(b_0 b_5) & \text{cov}(b_0 b_6) & \text{cov}(b_0 b_7) & \text{cov}(b_0 b_9) \\ \text{cov}(b_0 b_5) & \text{var}(b_5) & \text{cov}(b_5 b_6) & \text{cov}(b_5 b_7) & \text{cov}(b_5 b_9) \\ \text{cov}(b_0 b_6) & \text{cov}(b_5 b_6) & \text{var}(b_6) & \text{cov}(b_6 b_7) & \text{cov}(b_6 b_9) \\ \text{cov}(b_0 b_7) & \text{cov}(b_5 b_7) & \text{cov}(b_6 b_7) & \text{var}(b_7) & \text{cov}(b_7 b_9) \\ \text{cov}(b_0 b_9) & \text{cov}(b_5 b_9) & \text{cov}(b_6 b_9) & \text{cov}(b_7 b_9) & \text{var}(b_9) \end{bmatrix}$$

$$= x^{ii} S_b.$$

S_b is the variance–covariance matrix from the multivariate analysis of b_0, b_5, b_6, b_7 and b_9 and x^{ii} is the element of $(X^T X)^{-1}$ associated with P3/SR2 in the analysis of variance based on Equation (2.6). Here, because the design is orthogonal,

$$x^{ii} = 1 \bigg/ \sum x_i^2 \tag{2.15}$$

where the x_i are the values taken by P3/SR2 in Table 2.17.

Expressions (2.13) and (2.14) generalize to any linear function of any number of variables.

The complete set of predicted values, $\hat{c}1$–$\hat{c}27$, for the 27 weighing times and the variances and covariances of these predicted values are given by expressions (2.16) and (2.17) below:

$$c = T^T b$$
$$(27 \times 1) = (27 \times 5) \quad (5 \times 1) \tag{2.16}$$

$$\text{var}(c) = T^T V(b) T$$
$$(27 \times 27) \quad (27 \times 5) \quad (5 \times 5)(5 \times 27) \tag{2.17}$$

where c, T^T and b are matrices of the dimensions indicated.

In the example,

c		T^T			
		cos	sin	$Y1 \times \cos$	$Y2 \times \cos$
$c1$		1 1	0	1	1
$c2$		1 0.8781	0.4785	0.8781	0.8781
$c3$		1 0.1081	0.9941	0.1081	0.1081
$\vdots$	=	$\vdots$ $\vdots$	$\vdots$	$\vdots$	$\vdots$
$c25$		1 0.6960	-0.7181	0	-1.3920
$c26$		1 0.9495	-0.3139	0	-1.8989
$c27$		1 1	0	0	-2

b

$$
\begin{bmatrix}
1.0968 \\
0.2051 \\
-0.7371 \\
0.4332 \\
0.2696
\end{bmatrix}
=
\begin{bmatrix}
2.0047 \\
1.5413 \\
0.4622 \\
\vdots \\
1.3936 \\
1.0110 \\
0.7627
\end{bmatrix}
$$

Table 2.21 Analysis of variance of the values of b_0, b_1, b_2, b_3 and b_4 of Table 2.12

			b_0	
Source	*DF*	*MS*	*F*	*P(%)*
Blocks	2	0.715		
Stocking rate	1	385.150		
P1	1	11.894		
P1 × SR	1	3.198		
P2/SR1	1	39.166	7.6	1.5
P2/SR2	1	9.269	1.8	20.1
P3/SR1	1	35.299		
P3/SR2	1	16.489		
Error	14	5.137		

Variable	Regn coefficient	Standard error
P2/SR1	−2.5549	0.925
P2/SR2	1.2429	0.925

			b_3	
Source	*DF*	*MS*	*F*	*P(%)*
Blocks	2	0.012		
Stocking rate	1	108.568		
P1	1	0.046		
P1 × SR	1	46.303		
P2/SR1	1	0.041	0.00	96.2
P2/SR2	1	30.651	1.8	20.5
P3/SR1	1	90.423		
P3/SR2	1	0.355		
Error	14	17.319		

Variable	Regn coefficient	Standard error
P2/SR1	−0.0827	1.70
P2/SR2	−2.2602	1.70

where the columns of T^T were taken from Table 2.7 and the elements of b were taken from the P3/SR2 row of estimates in Table 2.23.

Var(c) is given in Table 2.25. The 5×5 matrix is the sums of squares and products matrix in Table 2.24 and the divisor, 14, is the number of degrees of freedom for error.

Expressions like (2.13) and (2.14) can only be readily evaluated by computer using matrix multiplication. However, special programs are not needed, as will be shown in section 2.9.

| | b_1 | | | | b_2 | |
MS	F	P(%)		MS	F	P(%)
9.624				15.488		
253.254				2.924		
335.212				88.548		
3.648				24.365		
0.131	0.00	95.1		19.647	1.2	28.4
68.647	2.0	17.8		92.348	5.8	3.0
94.280				34.367		
83.701				26.885		
34.115				15.852		
egn coefficient	Standard error			Regn coefficient	Standard error	
0.1480	2.38			1.8096	1.63	
3.3825	2.38			-3.9232	1.63	

| | b_4 | |
MS	F	P(%)
12.997		
40.619		
34.800		
10.838		
0.467	0.05	82.7
0.470	0.05	82.6
29.396		
35.157		
9.396		
egn coefficient	Standard error	
0.2791	1.25	
0.2798	1.25	

Table 2.22 Estimates of the coefficients of the regression of live weight on the values of $\cos d$, $\sin d$, $Y1 \times \cos d$ and $Y2 \times \cos d$ of Table 2.7 for each plot

B P SR	b_0	b_5 cos	b_6 sin	b_7 $Y1 \times \cos$	b_9 $Y2 \times \cos$
1 1 1	74.9297	−4.9840	−0.0222	−1.9061	0.6220
2 1 1	74.8621	−5.9714	−2.0804	−1.6273	0.9811
3 1 1	70.0449	−5.3064	−0.6879	−2.5049	1.2956
1 2 1	76.2044	−5.2569	0.1697	−0.5752	1.2699
2 2 1	74.1921	−5.8367	−0.1242	−1.2398	1.8836
3 2 1	78.6597	−6.1568	−0.2365	−1.3826	0.7278
1 1 2	68.0066	−5.8753	−2.3630	−2.1328	1.6567
2 1 2	67.9964	−3.6188	−2.9753	−0.4776	1.1765
3 1 2	64.1045	−4.5402	2.1242	−1.2416	1.4927
1 2 2	65.9981	−3.2087	0.5201	0.7294	2.3292
2 2 2	69.3472	−3.5256	−0.0490	−0.6643	1.5089
3 2 2	67.2454	−4.7083	−0.4874	−0.9923	1.5441
1 3 1	76.2520	−4.4143	−1.9308	−0.3780	1.9069
2 3 1	72.9475	−5.5473	−0.8540	−0.0865	2.5659
3 3 1	77.2574	−4.1871	−1.4567	0.1458	1.8322
1 4 1	81.3623	−3.3875	−2.8121	−0.0368	2.2965
2 4 1	81.4894	−4.5751	−2.1049	−0.9203	2.0522
3 4 1	78.8249	−5.5256	−1.1526	−0.2952	2.1051
1 3 2	69.6103	−5.9515	−1.9616	0.1164	1.1999
2 3 2	72.0271	−3.5282	−2.6423	−0.5636	2.2671
3 3 2	70.3368	−3.8362	−1.8219	−0.2475	2.7746
1 4 2	68.0857	−2.9613	−1.6216	1.6017	2.0724
2 4 2	65.7066	−2.7878	−2.5241	−0.3869	2.5694
3 4 2	70.0932	−3.9509	−1.5040	−0.1011	2.0597

2.9 CALCULATION OF PREDICTED VALUES AND THEIR VARIANCES BY REGRESSION ANALYSIS

2.9.1 P3/SR2

Consider Table 2.22 which contains the coefficients for the regression of live weight on $\cos d$, $\sin d$, $Y1 \times \cos d$ and $Y2 \times \cos d$ for each plot. Make up 27 new columns as follows:

$$(9) = (4) + 1 \times (5) + 0 \times (6) + 1 \times (7) + 1 \times (8)$$
$$(10) = (4) + 0.8781(5) + 0.4785(6) + 0.8781(7) + 0.8781(8)$$
$$\vdots$$
$$(34) = (4) + 0.9495(5) - 0.3139(6) + 0 \times (7) - 1.8989(8)$$
$$(35) = (4) + 0.90 \times (5) + 0 \times (6) \quad 0 \times (7) - 2 \times (8)$$

where the multipliers are again taken from Table 2.7.

Column 9 contains the predicted live weight for weighing 1 for each plot, column 10 contains the predicted live weight for weighing 2 for each plot, and so on. If the values in these columns are regressed on the values of the X-matrix of Table 2.17 the estimates obtained for the comparison P3/SR2 are the values of this comparison predicted from Equation (2.10). The variances obtained in the analyses are the correct variances for these predicted values. If multivariate regression is used, the covariances of the predicted values are also obtained but these are not usually required. The live weights predicted for each plot for weighings 1, 2 and 3 are given in Table 2.26 as examples.

The estimates of the comparisons obtained in the multivariate analysis of the predicted live weights for all 27 weighings are given in Table 2.27. Only the estimates for P3/SR2 are used because this was the only comparison whose values were fitted by $\cos d$, $\sin d$, $Y1 \times \cos d$ and $Y2 \times \cos d$. The standard errors of the predicted values are given by the standard deviation divided by $\sqrt{12}$. For example,

$$SE[\hat{c}_{5.2}(1)] = 2.4844/\sqrt{12}$$
$$= \pm 0.7172.$$

2.9.2 P3/SR1

Predicted values for $c_{5.1}$ were calculated by analysing live weights predicted for each plot from the estimates in Table 2.8. The predicted live weights for each plot for weighing 1, for example, were derived from columns 4 to 14 of Table 2.8 as

$$(15) = (4) - 0.3378(5) + 0.3954(6) - 0.3699(7) + 0.3488(8) + 1(9) + 0(10) + 1(11)$$
$$+ 0(12) + 1(13) + 0(14)$$

where the multipliers -0.3378, -0.3954, etc., came from the first row of Table 2.7 and ($\#$) denotes column $\#$.

The predicted values for the other weighings were calculated in the same way using the values in the successive rows of Table 2.7 as multipliers. The predicted values of the comparisons were obtained by analysing these predicted live weights. They are given in Table 2.28.

2.9.3 P1 and P2/SR2

Predicted values for c_3 and $c_{4.2}$, were calculated by analysing live weights predicted for each plot from the mean and the linear and quadratic coefficients in Table 2.14. The predicted live weights for each plot for

Table 2.23 Analyses of variance of b_0, b_5, b_6, b_7 and b_9 of Table 2.22

| | | | b_0 | |
Source	DF	MS	F	P(%)
Blocks	2	0.471		
Stocking rate	1	404.002		
P1	1	11.413		
P1 × SR	1	3.781		
P2/SR1	1	38.607		
P2/SR2	1	10.905		
P3/SR1	1	30.850		
P3/SR2	1	14.435	3.0	10.7
Error	14	4.859		

Variable	Regn coefficient	Standard error
P3/SR2	1.0968	0.636

| | | | b_7 | |
Source	DF	MS	F	P(%)
Source	2	0.587		
Stocking rate	1	1.732		
P1	1	2.770		
P1 × SR	1	0.001		
P2/SR1	1	0.145		
P2/SR2	1	0.545		
P3/SR1	1	4.896		
P3/SR2	1	2.252	5.5	3.4
Error	14	0.410		

Variable	Regn coefficient	Standard error
P3/SR2	0.4332	0.185

Table 2.24 Error sums of squares and products matrix for the multivariate analysis of the value of b_0, b_5, b_6, b_7 and b_9 of Table 2.22

b_0	b_5	b_6	b_7	b_9
68.021				
3.4398	11.826			
−16.215	−0.0045083	17.110		
−1.3827	2.6949	1.7001	5.7368	
−7.4328	1.6212	1.0142	−0.069664	3.0850

MS	b_5 F	P(%)	MS	b_6 F	P(%)
0.273			2.089		
6.674			0.169		
0.214			2.801		
1.068			0.030		
0.073			0.557		
2.179			0.100		
2.877			4.477		
0.505	0.60	45.2	6.520	5.3	3.7
0.845			1.222		

Regn coefficient	Standard error	Regn coefficient	Standard error
0.2051	0.265	−0.7371	0.319

MS	b_9 F	P(%)
0.090		
0.404		
0.346		
0.000		
0.004		
0.035		
2.979		
0.872	4.0	6.7
0.220		

Regn coefficient	Standard error
0.2696	0.136

weighing 1, for example, were derived from columns 4, 5 and 6 of Table 2.14 as:

$$(9) = (4) - 0.3378(5) + 0.3954(6)$$

where the multipliers −0.3378 and 0.3954 came from the first row of Table 2.7. The predicted values are given in Table 2.28.

2.9.4 P2/SR1

There was only one predicted value for P2/SR1, namely

$$\hat{c}_{4.1} = -2.5549 \text{ (Table 2.21)}$$
$$\text{var}(\hat{c}_{4.1}) = 5.137/6 \text{ (Table 2.21)}$$
$$= 0.856.$$

Table 2.25   Var (c) for $c_{5.2}$ (P3/SR2)

$$
\text{Var }(c)=
\begin{bmatrix}
1 & 1 & 0 & 1 & 1 \\
1 & 0.8781 & 0.4785 & 0.8781 & 0.8781 \\
1 & 0.1081 & 0.9941 & 0.1081 & 0.1081 \\
\cdots & & & & \\
1 & 0.6960 & -0.7181 & 0 & -1.3920 \\
1 & 0.9495 & -0.3139 & 0 & -1.8989 \\
1 & 1 & -0.009 & 0 & -2
\end{bmatrix}
\frac{1}{12 \times 14}
\begin{bmatrix}
68.021 & 3.4398 \\
3.4398 & 11.826 \\
-16.215 & 0.0045 \\
& \\
-1.3827 & 2.6949 \\
-7.4328 & 1.6212
\end{bmatrix}
$$

27×27 27×5

Table 2.26   Estimates of the coefficients of the within-plot regression of live weights on $\cos d$, $\sin d$, $Y1 \times \cos d$ and $Y2 \times \cos d$ and live weights predicted from these at weighings 1, 2 and 3

B P SR	b_0 (4)	cos (5)	sin (6)	$Y1 \times \cos$ (7)	$Y2 \times \cos$ (8)	P(1) (9)	P(2) (10)	P(3) (11)
			Coefficients of variables				*Predicted live weights*	
1 1 1	74.9297	−4.9840	−0.0222	−1.9061	0.6220	68.6616	69.4150	74.2300
2 1 1	74.8621	−5.9714	−2.0804	−1.6273	0.9811	68.2445	68.0557	72.0786
3 1 1	70.0449	−5.3064	−0.6879	−2.5049	1.2956	63.5292	63.9943	68.6567
1 2 1	76.2044	−5.2569	0.1697	−0.5752	1.2699	71.6422	72.2795	75.8799
2 2 1	74.1921	−5.8367	−0.1242	−1.2398	1.8836	68.9992	69.5728	73.5072
3 2 1	78.6597	−6.1568	−0.2365	−1.3826	0.7278	71.8481	72.5652	77.6882
1 1 2	68.0066	−5.8753	−2.3630	−2.1328	1.6567	61.6552	61.2987	64.9705
2 1 2	67.9964	−3.6188	−2.9753	−0.4776	1.1765	65.0765	64.0087	64.7230
3 1 2	64.1045	−4.5402	2.1240	−1.2416	1.4927	59.8154	61.3547	65.7525
1 2 2	65.9981	−3.2087	0.5201	0.7294	2.3292	65.8480	66.1151	66.4989
2 2 2	69.3472	−3.5256	−0.0490	−0.6643	1.5089	66.6662	66.9695	69.0086
3 2 2	67.2454	−4.7083	−0.4874	−0.9923	1.5441	63.0889	63.3624	66.3116
1 3 1	76.2520	−4.4143	−1.9308	−0.3780	1.9069	73.3666	72.7944	74.0207
2 3 1	72.9475	−5.5473	−0.8540	−0.0865	2.5659	69.8796	69.8449	71.7669
3 3 1	77.2574	−4.1871	−1.4567	0.1458	1.8322	75.0483	74.6205	75.5704
1 4 1	81.3623	−3.3875	−2.8121	−0.0368	2.2965	80.2345	79.0264	78.4449
2 4 1	81.4894	−4.5751	−2.1049	−0.9203	2.0522	78.0462	77.4587	79.0247
3 4 1	78.8249	−5.5256	−1.1526	−0.2952	2.1051	75.1092	75.0106	77.2774
1 3 2	69.6103	−5.5915	−1.9616	0.1164	1.1999	64.9751	64.6015	67.1592
2 3 2	72.0271	−3.5282	−2.6423	−0.5636	2.2671	70.2024	69.1605	69.2031
3 3 2	70.3368	−3.8362	−1.8219	−0.2475	2.7746	69.0277	68.3155	68.3841
1 4 2	68.0857	−2.9613	−1.6216	1.6017	2.0724	68.7985	67.9356	66.5507
2 4 2	65.7066	−2.7878	−2.5241	−0.3869	2.5694	65.1013	63.9673	63.1319
3 4 2	70.0932	−3.9509	−1.5040	−0.1011	2.0597	68.1009	67.6241	68.3827

$(9) = 4 + 1 \times (5) + 0 \times (6) + 1 \times (7) + 1 \times (8)$
$(10) = 4 + 0.8781(5) + 0.4785(6) + 0.8781(7) + 0.8781(8)$
$(11) = 4 + 0.1081(5) + 0.9941(6) + 0.1081(7) + 0.1081(8)$

$$\begin{bmatrix} -16.215 & -1.3827 & -7.43 \\ 0.0045 & 2.6949 & 1.62 \\ 17.001 & 1.7001 & 1.01 \\ \\ 1.7001 & 5.7368 & -0.06 \\ 1.0142 & -0.0697 & 3.08 \\ \\ 5 \times 5 \end{bmatrix} \times \begin{bmatrix} 1 & 1 & 1 & \cdots & 1 & 1 & 1 \\ 1 & 0.8781 & 0.1081 & \cdots & -0.6960 & -0.9495 & 1 \\ 0 & 0.4785 & 0.9941 & \cdots & -0.7181 & -0.3139 & 0 \\ \\ 1 & 0.8781 & 0.1081 & \cdots & 0 & 0 & 0 \\ 1 & 0.8781 & 0.1081 & \cdots & -1.3920 & -1.8989 & -2 \\ \\ & & & 5 \times 27 \end{bmatrix}$$

The standard error of $\hat{c}_{4.1}$ was thus $\pm\sqrt{0.856} = \pm 0.9252$ on 14 degrees of freedom. The rule that an estimate be rounded to one-tenth of its estimated standard error (Cochran and Cox, 1957; p. 60) leads to $\hat{c}_{4.1}$ being reported as -2.6 ± 0.925.

2.9.5 SR

Predicted values for c_6 were calculated by analysing the live weights predicted for each plot from the values of the coefficients in Table 2.12. The predicted live weights for each plot for weighing 1, for example, were derived from columns 4, 5, 6, 7, 8 and 9 as:

$$(10) = (4) - 0.3378(5) + 1(6) + 0(7) + 0(8) + 0(9)$$

where the multipliers $-0.3378, 1, 0, 0$ and 0 came from the first row of Table 2.7.

The predicted values for all comparisons and their standard errors are given in Table 2.28 together with the 95% confidence limits for $\hat{c}_{5.2}$ as an example. The fitted regressions are shown in Figure 2.4.

2.10 CONFIDENCE LIMITS

Confidence limits of $(100 - \alpha)\%$ may be calculated for the predicted values as $c \pm st$ where t is the value of Student's t-distribution with n_e degrees of freedom and probability level $\alpha\%$, s is the standard error of c, and n_e is the number of degrees of freedom for error.

For example, using 2.145 as the 5% critical value for the t-distribution on 14 degrees of freedom, the 95% confidence limits for $c1$, the predicted value for $c_{5.2}$ at the first weighting are

$$c1(L), \ c1(U) = 2.0046 \pm 0.717 \times 2.145$$
$$= 2.0046 \pm 1.5380$$
$$= 0.467, \ 3.543$$

where L is lower and U is upper.

Table 2.27   Predicted values for $c_{5.2}$ (P3/SR2) for 27 weighing times and standard

	1	2	3	4	5	6
Factor						
Constant term	68.8735	68.7230	70.7593	74.1322	75.9047	72.7074
Stocking rate	3.1772	3.3302	4.0862	4.8565	4.8719	3.8140
P1	1.7592	1.8948	1.5402	0.7005	0.0553	0.6797
P1 × SR	0.2498	0.2640	0.4781	0.7637	0.8580	0.5403
P2/SR1	−2.5159	−2.3726	−2.2315	−2.3524	−2.7024	−2.8290
P2/SR2	0.3674	0.4251	1.1135	2.0034	2.2686	1.2548
P3/SR1	3.2300	2.7394	1.1720	−0.0249	0.4767	2.5626
P3/SR2	2.0046	1.5413	0.4622	−0.0720	0.6600	2.0183
	2.4844	2.2636	1.9356	2.3319	2.9492	2.8505

*Predicted values are calculated for all contrasts, but those for P3/SR2 are the only

Table 2.27   Cont.

	15	16	17	18	19	20
Factor						
Constant term	73.6212	71.5659	70.1878	64.2243	65.7555	71.4497
Stocking rate	4.0396	3.7642	3.7113	3.8695	3.9726	4.1974
P1	0.5021	0.5134	0.7787	0.9137	1.1708	1.4662
P1 × SR	0.6269	0.3809	0.2661	0.2729	0.3238	0.5230
P2/SR1	−2.8255	−2.9572	−2.8393	−2.5087	−2.3640	−2.2298
P2/SR2	1.4973	1.1645	0.9758	0.8806	0.9389	1.2369
P3/SR1	2.1949	2.2779	1.9771	0.9447	0.8131	1.0134
P3/SR2	1.8307	1.6364	1.1683	0.5683	0.3394	0.3737
	2.9000	2.7420	2.4323	2.8098	2.4837	1.9165

*Predicted values are calculated for all contrasts, but those for P3/SR2 are the only

deviations used to calculate their standard errors*

Weighing time							
7	8	9	10	11	12	13	14
Predicted value							
71.7702	69.0466	69.8579	70.1937	71.9934	73.2109	74.4570	72.5086
3.5982	3.1739	3.7685	3.9810	4.3216	4.4537	4.4936	4.1205
0.9121	1.6937	0.9788	1.3048	1.4896	1.4222	1.1669	0.9295
0.4606	0.2584	0.2653	0.3692	0.6225	0.7564	0.8572	0.6893
−2.7967	−2.5536	−2.7019	−2.4183	−2.1475	−2.1181	−2.2370	−2.5122
1.0066	0.3908	0.9488	1.0469	1.3643	1.5533	1.7219	1.4346
2.9005	3.2929	1.7163	1.2495	0.9077	0.9425	1.2358	1.8850
2.1715	2.0886	0.8773	0.4795	0.3947	0.5894	1.0328	1.5632
Standard deviations							
2.7929	2.5304	2.2324	1.9683	2.0257	2.2321	2.5485	2.7370

appropriate ones.

Weighing time						
21	22	23	24	25	26	27
Predicted value						
80.4847	80.2336	77.5891	71.1619	67.5450	65.0142	64.2449
4.3524	4.3177	4.1728	3.9463	3.8561	3.8221	3.8339
1.0954	0.9789	0.6733	0.4534	0.4848	0.6278	0.7645
0.8517	0.8445	0.7545	0.5279	0.3981	0.3054	0.2758
−2.5294	−2.5985	−2.7637	−2.8459	−2.7975	−2.6897	−2.6001
1.8130	1.8126	1.6949	1.3445	1.1283	0.9621	0.8999
2.2077	2.3067	2.3896	2.0451	1.6892	1.3139	1.1029
1.5526	1.6895	1.8993	1.7086	1.3936	1.0110	0.7701
Standard deviations						
1.9731	2.1201	2.5599	3.0210	3.1191	3.0682	2.9597

appropriate ones.

Table 2.28 Predicted values and standard errors of the predicted values for the regression coefficients of Equation (2.6) and 95% confidence limits for $c_{5.2}$

Weighing	c_6 (SR)	SE	c_3 (P1)	SE	$c_{4.1}$ (P2/SR1)	SE	$c_{4.2}$ (P2/SR2)	SE	$c_{5.1}$ (P3/SR1)	SE	$c_{5.2}$ (P3/SR2)	SE	95% limits for $c_{5.2}$
1	2.44	0.622	1.71	0.929	−2.55	0.925	−1.45	1.313	3.48	0.824	2.00	0.717	0.472 3.548
2	2.63	0.533	1.78	0.868			−1.16	1.228	2.98	0.809	1.54	0.634	0.137 2.943
3	3.16	0.429	1.90	0.767			−0.63	1.084	1.83	0.717	0.46	0.559	−0.739 1.659
4	3.58	0.360	1.90	0.685			−0.14	0.969	1.09	0.720	−0.07	0.673	−1.516 1.376
5	3.57	0.437	2.06	0.612			0.41	0.866	1.71	0.969	0.66	0.851	−1.165 2.485
6	3.16	0.614	2.07	0.570			0.90	0.806	3.46	1.046	2.02	0.822	0.257 3.783
7	3.10	0.615	2.07	0.565			0.99	0.799	3.74	1.035	2.17	0.806	0.441 3.899
8	3.10	0.532	2.04	0.558			1.26	0.789	4.10	0.966	2.09	0.730	0.524 3.656
9	3.80	0.478	2.01	0.559			1.43	0.790	2.94	0.629	0.88	0.644	−0.504 2.264
10	4.92	0.436	1.95	0.565			1.60	0.799	1.63	0.602	0.48	0.568	−0.738 1.698
11	5.50	0.400	1.86	0.579			1.79	0.818	0.67	0.628	0.39	0.585	−0.865 1.645
12	5.25	0.390	1.79	0.588			1.88	0.832	0.70	0.665	0.59	0.644	−0.791 1.971
13	4.41	0.450	1.69	0.603			1.98	0.853	1.30	0.731	1.03	0.735	−0.547 2.607
14	3.46	0.565	1.58	0.618			2.05	0.875	2.07	0.793	1.56	0.787	−0.254 3.174
15	2.47	0.695	1.39	0.644			2.13	0.911	2.83	0.849	1.83	0.837	0.035 3.625
16	2.75	0.655	1.18	0.671			2.16	0.949	2.25	0.835	1.64	0.792	−0.057 3.337
17	3.79	0.550	0.99	0.695			2.15	0.983	0.85	0.803	1.17	0.702	−0.336 2.676
18	3.74	0.559	0.90	0.707			2.14	1.000	−1.90	1.245	0.57	0.811	−1.170 2.310
19	3.63	0.637	0.76	0.727			2.12	1.028	−1.07	1.164	0.34	0.716	−1.196 1.876
20	3.86	0.757	0.49	0.763			2.04	1.079	0.69	1.128	0.37	0.553	−0.804 1.564
21	5.20	0.649	−0.12	0.856			1.79	1.211	1.78	0.988	1.55	0.570	0.339 2.781
22	5.34	0.650	−0.21	0.871			1.74	1.231	1.56	0.992	1.69	0.612	0.377 3.003
23	5.53	0.696	−0.46	0.914			1.61	1.292	0.80	1.090	1.90	0.739	0.315 3.485
24	5.35	0.726	−0.82	0.981			1.39	1.387	0.32	1.195	1.71	0.872	−0.165 3.585
25	5.07	0.698	−1.04	1.024			1.26	1.448	0.83	1.106	1.39	0.900	−0.543 3.323
26	4.76	0.674	−1.27	1.070			1.10	1.513	2.17	0.932	1.01	0.886	−0.890 2.910
27	4.58	0.685	−1.42	1.102	−2.55	0.925	1.00	1.558	3.48	0.875	0.77	0.854	−1.072 2.592

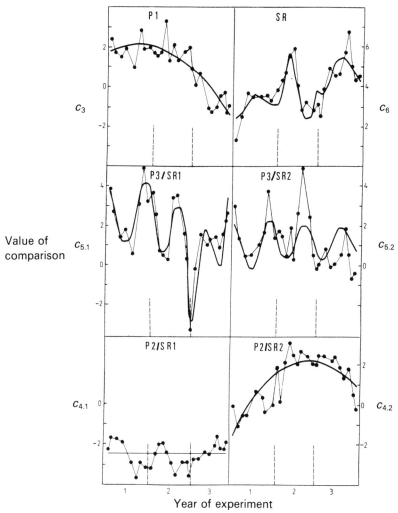

Figure 2.4 Observed values and fitted regressions for six treatment comparisons.

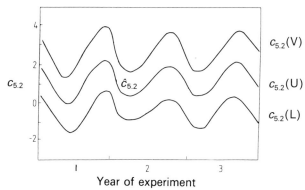

Figure 2.5 Regression of the values of $c_{5.2}$ on time and 95% confidence boundaries for the regression.

The set of such limits for the 27 weighings provides confidence boundaries for the predicted values as shown in Figure 2.5.

2.11 DISCUSSION

The procedure used in this analysis was certainly long. However, a lot has been accomplished. Analyses of variance have been carried out on data from three replicates of a 4×2 factorial experiment measured on 27 occasions. Regressions have been fitted to six sets of 27 values of treatment comparisons to combine the information available over three years. The values of each treatment comparison have been smoothed to allow general conclusions. Trends in the values have been tested for significance.

Different regression equations could well have been used. For example, the values of c_3 could probably be fitted by three regressions, one for each year. During the first year, the values of c_3 were fairly constant; they decreased slightly throughout year two, and markedly in year three. The choice of a regression function is not critical provided it performs its functions of smoothing the individual values and providing tests of trend.

The observed values of $c_{5.2}$ deviate markedly and systematically from the fitted regression. This is particularly so towards the end of each year. This comparison is considered further in Chapter 7.

The agronomic interpretation of the analyses, including the regressions, requires information on rainfall and on the composition of the pasture which is not given here. For example, the comparison P1 is the difference between sheep on pastures sown to clover and phalaris (SP) and those only sown to clover (S). However, the clover pastures also contained grasses already present in the area. These were mainly rye grass and barley grass. Barley grass is considered to have lower quality than rye grass but lasts longer into summer. Rye grass was the main grass in the clover pasture in the first two years, but was replaced by barley grass in the last year. The last year was also drier than average. Sowing phalaris with the clover produced heavier sheep during the first two years, but not in the third year.

3 A 3 × 3 factorial with quantitative levels

3.1 INTRODUCTION

The regressions in time fitted in Chapters 1 and 2 described the trends in the values of plasma thyroxine in sheep and in the differences in live weights between sheep on different pastures. The regressions could be interpreted directly in terms of the measured response. In more complex experiments such as factorials, however, the regressions in time are often fitted to treatment effects or regression coefficients and are not directly interpretable in terms of the measured response variable; a further step is needed. The predicted values of the effects are used to calculate predicted values of the response. This will be demonstrated in this chapter.

3.2 BACKGROUND TO THE EXPERIMENT

Brownlee *et al.* (1975) describe a grazing experiment on dryland lucerne conducted at Condobolin, New South Wales. The aim was 'to determine whether annual topdressings of superphosphate on dryland lucerne increase sheep and plant production...'. Two randomized blocks were used to compare the nine treatments from the factorial combinations of stocking rates (S) at 3.7, 4.9 and 6.2 dry ewes per hectare and annual applications of superphosphate (P) at 0, 125 and 250 kg per hectare. The lucerne was sown in 1968 and the ewes were put on the plots in August 1969. The experiment was ended in December 1972 because of drought.

The sheep were weighed about every 40 days from 29 March 1971 until 28 December 1972, a total of 17 weighings as follows:

1971 – 29 March, 7 May, 17 June, 6 August, 6 September, 15 October, 24 November;

1972 – 4 January, 10 February, 23 March, 2 May, 11 June, 21 July, 30 August, 9 October, 17 November, 28 December.

These were taken to be equally spaced time-points 1, 2,..., 17.

Available forage as dry matter was estimated for each period. The nitrogen and phosphorus content of the lucerne was also estimated for each period. Live weights will be considered in this chapter and the relations between live weight and dry matter, and between nitrogen and phosphorus content, will be considered in Chapter 9. The experiment was badly affected by drought and the sheep on some treatments were hand fed at certain periods. The details are given in Brownlee *et al.* (1975). This probably accounts for certain of the trends in the coefficients. From May 1972, hand feeding would have progressively eliminated effects of S and reduced effects of P at 4.9 and at 6.3 ewes per hectare.

3.3 DATA AND ANALYSIS OF VARIANCE

The live weights are given in Table 3.1 These were transformed to logarithms before analysis to make the variances more constant over time. The analysis of variance, given in detail in Table 3.2, follows the usual form for a 3 × 3 factorial with quantitative factors.

3.3.1 Estimates of the treatment effects at weighing 5

The live weights at weighing 5 will be used to illustrate these treatment effects. The live weights in natural logarithms are set out in Table 3.3. The linear and quadratic coefficients for the regression of live weight on the levels of superphosphate can be calculated directly by using the sets of orthogonal polynomial coefficients $(-1, 0, 1)$ and $(1, -2, 1)$ to give:

$$\text{mean linear regression on P denoted by } P = (-1 \times 21.328 + 0 \times 21.663$$
$$+ 1 \times 21.930)/2 \times 6$$
$$= 0.602/12,$$

$$\text{mean quadratic regression on } P(P^2) = (21.328 - 2 \times 21.663$$
$$+ 21.930)/6 \times 6$$
$$= -0.068/36.$$

The divisors in these calculations are the sums of squares of the coefficients multiplied by 6, which is the number of observations in the totals.
    Similarly,

$$S = (-22.237 + 21.105)/12$$
$$= -1.132/12,$$

and

$$S^2 = (22.237 - 2 \times 21.579 + 21.105)/36$$
$$= 0.184/36.$$

The components of the interaction of P and S are simplified by first calculating the linear and quadratic regressions of live weight on the levels

Table 3.1 Live weights of ewes (kg) grazing lucerne at three levels of superphosphate (P) and three rates of stocking (S) in two blocks (B) on 17 occasions

B S P			1	2	3	4	5	6	7	8	9	10	11	12	13	14	15	16	17
1	1	1	47.7	49.1	44.0	37.4	39.3	46.0	43.0	49.5	47.2	50.1	45.9	41.4	39.4	36.9	36.9	35.7	34.8
1	1	2	46.9	46.6	41.8	35.8	39.1	46.7	46.2	50.1	48.1	51.1	48.8	43.0	41.9	40.5	38.8	36.6	36.0
1	1	3	49.7	50.2	45.0	39.0	44.2	48.9	48.9	53.3	54.2	53.7	51.0	43.9	45.0	40.7	40.4	34.8	37.1
1	2	1	45.8	44.7	39.1	32.9	35.4	40.3	38.3	44.1	45.3	45.1	40.9	35.4	36.4	35.4	35.6	33.3	34.9
1	2	2	48.4	47.8	41.3	34.3	37.0	42.5	41.2	47.9	48.1	50.4	46.6	38.1	39.7	37.7	37.7	35.6	36.8
1	2	3	47.2	46.9	40.9	34.5	37.1	42.8	41.4	47.6	48.4	50.2	44.9	36.9	38.9	37.6	39.1	36.4	38.1
1	3	1	43.9	42.9	35.1	33.1	32.9	38.4	36.3	41.7	42.2	41.9	39.6	34.4	35.1	34.1	34.2	33.2	34.0
1	3	2	46.5	44.0	37.4	33.2	32.0	39.1	38.0	44.0	44.9	44.3	40.7	34.9	36.6	35.9	36.2	34.7	37.6
1	3	3	45.8	43.5	37.9	33.0	34.0	38.7	37.8	42.7	44.9	43.4	41.6	35.1	37.1	35.5	36.6	33.6	35.8
2	1	1	45.5	44.2	41.3	34.6	35.5	41.8	41.0	46.0	48.0	48.1	47.4	39.8	39.8	37.1	37.9	35.3	33.7
2	1	2	48.6	48.4	44.5	39.4	41.3	47.3	46.4	49.4	50.5	50.0	47.7	42.3	41.5	40.5	39.7	35.4	36.7
2	1	3	49.8	49.8	46.5	41.1	45.6	48.5	48.3	49.8	49.2	50.9	49.7	43.5	45.0	42.0	41.9	38.9	40.5
2	2	1	46.0	45.1	39.8	33.4	35.0	40.5	38.8	43.7	43.2	47.5	44.8	36.9	37.5	36.8	36.9	35.4	34.9
2	2	2	48.0	46.5	42.4	35.2	37.8	43.5	42.9	48.6	48.6	49.0	46.7	38.3	40.1	38.5	39.3	37.2	38.4
2	2	3	45.7	47.0	40.0	33.4	36.6	44.8	40.3	48.9	49.1	50.4	44.7	39.0	38.2	38.4	36.9	36.7	36.7
2	3	1	43.1	43.7	35.3	31.4	32.2	37.3	33.6	41.2	41.9	42.8	36.2	32.3	33.7	33.7	33.4	33.8	35.7
2	3	2	44.2	44.8	38.0	33.0	35.4	42.4	38.5	45.4	45.4	48.3	42.0	36.3	38.7	38.0	37.7	37.7	38.5
2	3	3	46.3	46.0	39.9	34.0	35.9	41.3	38.7	45.5	47.0	47.9	42.0	37.3	38.0	37.5	37.0	36.0	36.6

Table 3.2

Source		DF
Blocks,	B	1
Mean linear regression on P (linear effect of P),	P	1
Mean quadratic regression on P (quadratic effect of P),	P^2	1
Linear effect of S,	S	1
Quadratic effect of S,	S^2	1
Interactions of P and S,	$P \times S$	1
	$P \times S^2$	1
	$P^2 \times S$	1
	$P^2 \times S^2$	1
Error		8
Total		17

Table 3.3 Live weights at weighing 5 (natural logarithms)

S (ewes/ha)	Block	P (kg/ha) 0	125	250	Totals
3.7	1	3.671	3.666	3.789	
	2	3.570	3.721	3.820	
Sub-totals		7.241	7.387	7.609	22.237
4.9	1	3.567	3.611	3.614	
	2	3.555	3.632	3.600	
Sub-totals		7.122	7.243	7.214	21.579
6.3	1	3.493	3.466	3.526	
	2	3.472	3.567	3.581	
Sub-totals		6.965	7.033	7.107	21.105
Totals		21.328	21.663	21.930	64.921

of P for each level of S separately as follows:

$$P/S\ 3.7 = (-7.241 + 7.609)/2 \times 2 = 0.368/4,$$
$$P/S\ 4.9 = (-7.122 + 7.214)/4 = 0.092/4,$$
$$P/S\ 6.3 = (-6.965 + 7.107)/4 = 0.142/4,$$
$$P^2/S\ 3.7 = (7.241 - 2 \times 7.387 + 7.609)/6 \times 2 = 0.076/12,$$
$$P^2/S\ 4.9 = (7.122 - 2 \times 7.243 + 7.214)/12 = -0.15/12,$$
$$P^2/S\ 6.3 = (6.965 - 2 \times 7.033 + 7.107)/12 = 0.006/12.$$

Then,

$P \times S$ = the linear coefficient of the regression of the values of P/S on the levels of S

$$= (-1 \times 0.368 + 0 \times 0.092 + 1 \times 0.142)/4 \times 2$$
$$= -0.226/8$$

$P \times S^2$ $=$ the quadratic coefficient of the regression of the values of P/S
on S
$$= (1 \times 0.368 - 2 \times 0.092 + 1 \times 0.142)/4 \times 6$$
$$= 0.326/24$$

$P^2 \times S$ $=$ the linear coefficient of the regression of the values of P^2/S on S
$$= (-1 \times 0.076 + 0 \times -0.15 + 1 \times 0.006)/12 \times 4$$
$$= -0.070/48$$

$P^2 \times S^2$ $=$ the quadratic coefficient of the regression of the values of P^2/S
on S
$$= (0.076 - 2 \times -0.15 + 0.006)/12 \times 6$$
$$= 0.382/72.$$

The analysis of variance can be carried out as usual by partitioning the sum of squares of the observations. However, it is more conveniently obtained when a large number of analyses are to be carried out by fitting the linear model:

$$\hat{y} = c_0 + c_1 x_1 + c_2 x_2 + c_3 x_3 + c_{22} x_2^2 + c_{33} x_3^2 + c_{23} x_2 x_3 + c_{233} x_2 x_3^2$$
$$+ c_{223} x_2^2 x_3 + c_{2233} x_2^2 x_3^2 \qquad (3.1)$$

where, x_1 is a block variable taking the values 1 for block 1 and -1 for block 2, x_2 and x_3 are linear treatment contrasts $(-1, 0, 1)$ for P and S, x_2^2 and x_3^2 are quadratic treatment contrasts $(1, -2, 1)$ for P and S, c_0 is the constant term, and $c_2, \ldots, c_{2233}$ are main and interaction factorial effects of S and P.

The design matrix, X, is given in Table 3.4. For this analysis $(X^T X)$ was the following matrix:

$$\begin{bmatrix} 18 & 0 & 0 & 0 & 0 & 0 & 0 & 0 \\ 0 & 18 & 0 & 0 & 0 & 0 & 0 & 0 \\ 0 & 0 & 12 & 0 & 0 & 0 & 0 & 0 \\ 0 & 0 & 0 & 36 & 0 & 0 & 0 & 0 \\ 0 & 0 & 0 & 0 & 8 & 0 & 0 & 0 \\ 0 & 0 & 0 & 0 & 0 & 24 & 0 & 0 \\ 0 & 0 & 0 & 0 & 0 & 0 & 24 & 0 \\ 0 & 0 & 0 & 0 & 0 & 0 & 0 & 72 \end{bmatrix}.$$

This type of matrix is called diagonal because the only non-zero elements occur on the diagonal.

Table 3.4 Design matrix used to fit Equation (3.1)

Block	S	P	Mean x_0	B x_1	P x_2	P² x_2^2	S x_3	S² x_3^2	P×S x_2x_3	P×S² $x_2x_3^2$	P²×S $x_2^2x_3$	P²×S² $x_2^2x_3^2$
1	1	1	1	1	-1	1	-1	1	1	-1	-1	1
1		2	1	1	0	-2	-1	1	0	0	2	-2
1		3	1	1	1	1	-1	1	-1	1	-1	1
1	2	1	1	1	-1	1	0	-2	0	2	0	-2
1		2	1	1	0	-2	0	-2	0	0	0	4
1		3	1	1	1	1	0	-2	0	-2	0	-2
1	3	1	1	1	-1	1	1	1	-1	-1	1	1
1		2	1	1	0	-2	1	1	0	0	-2	-2
1		3	1	1	1	1	1	1	1	1	1	1
2	1	1	1	-1	-1	1	-1	1	1	-1	-1	1
2		2	1	-1	0	-2	-1	1	0	0	2	-2
2		3	1	-1	1	1	-1	1	-1	1	-1	1
2	2	1	1	-1	-1	1	0	-2	0	2	0	-2
2		2	1	-1	0	-2	0	-2	0	0	0	4
2		3	1	-1	1	1	0	-2	0	-2	0	-2
2	3	1	1	-1	-1	1	1	1	-1	-1	1	1
2		2	1	-1	0	-2	1	1	0	0	-2	-2
2		3	1	-1	1	1	1	1	1	1	1	1
Sum of squares			18	18	12	36	12	36	8	24	24	72

It is written:

$$\text{diag.}(18, 18, 12, 36, 8, 24, 24, 72).$$

The elements are the sums of squares given in Table 3.4. Then

$$(X^T X)^{-1} = \text{diag.}(1/18, 1/18, 1/12, 1/36, 1/8, 1/24, 1/24, 1/72).$$

$(X^T X)$ was diagonal because the columns of X were mutually orthogonal.

3.3.2 Predicted values for live weight at weighing 5

The detailed analysis carried out later indicated that all variables except $P^2 \times S$ and $P^2 \times S^2$ were important at weighing 5. Equation (3.1) was then reduced to

$$\hat{y} = c_0 + c_1 x_1 + c_2 x_2 + c_{22} x_2^2 + c_3 x_3 + c_{33} x_3^2 + c_{23} x_2 x_3 + c_{233} x_2 x_3^2$$

to calculate the predicted values $\hat{y}$. These are calculated from

$$Y = X\beta$$

where X only contains the values of those treatment variables that have been found to be important and β contains the estimates of their effects. Specifically,

$$X \qquad\qquad\qquad\qquad \beta$$

c_0	x_2	x_2^2	x_3	x_3^2	x_2x_3	$x_2x_3^2$
(mean)	(P)	(P²)	(S)	(S²)	(P × S)	(P × S²)

$$
Y = \begin{bmatrix}
1 & -1 & 1 & -1 & 1 & 1 & -1 \\
1 & 0 & -2 & -1 & 1 & 0 & 0 \\
1 & 1 & 1 & -1 & 1 & -1 & 1 \\
1 & -1 & 1 & 0 & -2 & 0 & 2 \\
1 & 0 & -2 & 0 & -2 & 0 & 0 \\
1 & 1 & 1 & 0 & -2 & 0 & -2 \\
1 & -1 & 1 & 1 & 1 & -1 & -1 \\
1 & 0 & -2 & 1 & 1 & 0 & 0 \\
1 & 1 & 1 & 1 & 1 & 1 & 1
\end{bmatrix}
\begin{bmatrix}
64.921/18 \\
0.602/12 \\
-0.068/36 \\
-1.132/12 \\
0.184/36 \\
-0.226/8 \\
0.326/24
\end{bmatrix}
$$

$$
= \begin{bmatrix}
3.612 \\
3.710 \\
3.796 \\
3.572 \\
3.600 \\
3.618 \\
3.480 \\
3.521 \\
3.511
\end{bmatrix}.
$$

The mean will often be left out of this type of calculation as it affects each predicted value equally.

3.4 ESTIMATES OF $c_2,\ldots,c_{2233}$ OF EQUATION (3.1) FOR EACH WEIGHING TIME

Equation (3.1) was fitted to the natural logarithms of the live weights at each weighing. The estimates of $c_2,\ldots,c_{2233}$ for each weighing time are given in Table 3.5 and are plotted against time in Figure 3.1. The scales of the ordinates in Figure 3.1 were chosen to indicate the average contribution of each coefficient to the predicted live weight based on Equation (3.1) and the values of the treatment variables in Table 3.4.

Table 3.5   Estimates of the mean and the main and interaction effects of Equation (3.1) for 17

Variable		1	2	3	4	5	6	7	Weighing 8 (estimates except
constant	c_0	3.8411	3.8314	3.7000	3.5507	3.6067	3.7538	3.7109	3.8399
P	c_2	22.5944	24.6615	32.5272	28.1716	50.0949	40.0457	49.1670	38.7593
P^2	c_{22}	−5.3415	−2.1304	−4.6314	−3.5577	−1.8797	−9.4041	−14.2136	−10.5831
S	c_3	−32.7553	−41.9745	−81.4070	−69.0452	−94.2826	−81.3425	−102.4603	−67.3321
S^2	c_{33}	−2.8081	−2.0907	−1.4944	13.0446	5.0960	3.6896	5.3714	−2.4785
$P \times S$	$c_{2,3}$	−2.4485	−9.5330	7.2726	−17.1898	−28.2770	−12.5191	−13.8291	−3.7957
$P \times S^2$	$c_{2,33}$	8.3513	1.1671	0.0111	8.1500	13.5968	−0.1138	10.1132	−4.2206
$P^2 \times S$	$c_{22,3}$	−3.4447	−4.1495	−7.0624	−2.9405	−2.8241	−4.7983	−3.4758	−6.8823
$P^2 \times S^2$	$c_{22,33}$	4.4953	3.4487	5.4353	4.0894	5.3405	−1.0300	2.5231	2.5860

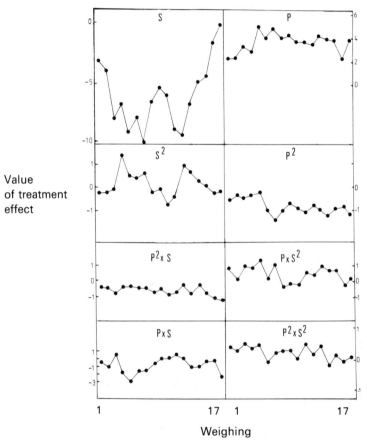

Value of treatment effect

Weighing

Figure 3.1   Values of the factorial main and interaction effects of Equation (3.1) at 17 weighing times.

weighings

9 $c_0 \times 10^3$)	10	11	12	13	14	15	16	17
3.8484	3.8702	3.7920	3.6408	3.6617	3.6253	3.6246	3.5707	3.5959
44.5058	36.6765	36.8081	34.0373	43.0097	39.3333	37.9095	22.6407	38.5255
−6.8126	−8.7354	−10.8190	−7.5854	−9.9608	−12.4674	−9.3009	−8.9809	−11.8200
−54.7653	−62.2143	−91.5020	−94.4323	−70.7623	−50.7782	−45.5486	−18.2458	−0.9300
−1.5877	−8.0183	−4.2515	9.3808	6.2165	2.0040	−0.6677	−2.8522	−2.1539
1.7375	2.7223	5.6423	2.0680	−10.0235	−9.3738	−2.5960	0.4320	−21.4149
−2.0235	−2.4610	7.0200	4.9124	10.8899	6.8102	7.2327	−4.3141	2.0106
−5.4695	−8.9332	−7.1555	−2.0058	−9.7447	−2.3540	−7.6427	−10.3279	−11.8858
3.3147	0.5492	5.1297	1.4003	4.2994	−1.5101	1.4695	−0.0032	0.7211

3.5 REGRESSION OF $c_2, \ldots, c_{2233}$ ON TIME

No more than eight coefficients could be used to describe the regression of $c_2, \ldots, c_{2233}$ on time because there were only eight degrees of freedom for error.

The following equation was used to fit polynomial trends in time to the estimates of $c_2, \ldots, c_{2233}$:

$$\hat{c} = b_0 + b_1\xi_1 + b_2\xi_2 + b_3\xi_3 + b_4\xi_4 + b_5\xi_5 + b_6\xi_6 + b_7\xi_7.$$

The ξ_i are the orthogonal polynomials replacing $t, t^2, \ldots, t^7$. As usual, the trends were fitted in two stages.

1. The following equation was fitted to the live weights of each plot:

$$\hat{y} = b_0 + b_1\xi_1 + b_2\xi_2 + b_3\xi_3 + b_4\xi_4 + b_5\xi_5 + b_6\xi_6 + b_7\xi_7. \qquad (3.2)$$

2. Equation (3.1) was fitted to the estimates of $b_0, b_1, \ldots, b_7$ for each plot to estimate $b_0, b_1, \ldots, b_7$ for each of the sequences $c_2, c_{22}, \ldots, c_{2233}$ and to test their significance.

3.6 FITTING EQUATION (3.2) TO THE LIVE WEIGHTS FOR EACH PLOT

The live weights for each plot can be formally regressed on the values of $\xi_1, \ldots, \xi_7$ to obtain the estimates of the b_k of Equation (3.2) for each plot. The values of the ξ_i and the live weights for plots 1 and 18 (Table 3.1) as examples are given in Table 3.6 in the usual layout for regression analysis. However, the b_k can be calculated directly from the data in Table 3.1 because the ξ_i are orthogonal. The data are first transformed to natural

Table 3.6 Live weights for plots 1 and 18 and orthogonal variables used to fit seventh degree polynomial to 17 equally spaced weighings

Time	Live weight Plot 1	Plot 18	ξ_0	ξ_1	ξ_2	ξ_3	ξ_4	ξ_5	ξ_6	ξ_7
1	47.7	46.3	1	-8	40	-28	52	-104	104	-130
2	49.1	46.0	1	-7	25	-7	-13	91	-169	325
3	44.0	39.9	1	-6	12	7	-39	104	-78	-39
4	37.4	34.0	1	-5	1	15	-39	39	65	-247
5	39.3	35.9	1	-4	-8	18	-24	-36	128	-149
6	46.0	41.3	1	-3	-15	17	-3	-83	93	75
7	43.0	38.7	1	-2	-20	13	17	-88	2	215
8	49.5	45.5	1	-1	-23	7	31	-55	-85	175
9	47.2	47.0	1	0	-24	0	36	0	-120	0
10	50.1	47.9	1	1	-23	-7	31	55	-85	-175
11	45.9	42.0	1	2	-20	-13	17	88	2	-215
12	41.4	37.3	1	3	-15	-17	-3	83	93	-75
13	39.4	38.0	1	4	-8	-18	-24	36	128	149
14	36.9	37.5	1	5	1	-15	-39	-39	65	247
15	36.8	37.0	1	6	12	-7	-39	-104	-78	39
16	35.7	36.0	1	7	25	7	-13	-91	-169	-325
17	34.8	36.6	1	8	40	28	52	104	104	130
Sum of squares			17	408	7752	3876	1679	100776	178296	579462

logarithms. Then, writing ($\#$) to indicate column $\#$,

$$b_0 = [(4) + (5) + \cdots + (19) + (20)]/17,$$
$$b_1 = [(4) \times -8 + (5) \times -7 + \cdots + (19) \times 7 + (20) \times 8]/408,$$
$$b_2 = [(4) \times 40 + (5) \times 25 + \cdots + (19) \times 25 + (20) \times 40]/7752;$$

and so on; the multipliers were taken from the columns of Table 3.6 and the divisors are their sums of squares. This is easily accomplished as most statistical programs have facilities for deriving new columns of data from existing ones. These derived values are given in Table 3.7 and become the data for the next step.

3.7 ANALYSIS OF THE b_k OF TABLE 3.7

The values of the b_k were regressed in turn on the values of the x-variables of the design matrix (Table 3.4) to estimate the coefficients $b_0, \ldots, b_7$ for the regressions of the treatment effects on time and to test their significance. The analyses are given in Table 3.8.

A summary table, such as Table 3.9, together with Figure 3.1, is helpful in interpreting these analyses.

Table 3.7 Polynomial coefficients of order 0 to 7 for each plot; the coefficients of order 1 to 7 have been multiplied by 10^5

			Column				
(4)	(5)	(6)	(7)	(8)	(9)	(10)	(11)
			polynomial coefficient of order				
B S P 0	1	2	3	4	5	6	7
1 1 1 3.745	−1480.411	−216.974	−227.387	174.440	12.690	−37.600	7.687
1 1 2 3.764	−852.635	−282.255	−286.835	184.983	−1.040	−27.991	12.589
1 1 3 3.818	−1347.536	−339.252	−250.906	190.344	8.225	−22.653	15.155
1 2 1 3.657	−1116.558	−102.545	−243.082	209.033	2.656	−38.990	12.186
1 2 2 3.726	−999.629	−151.418	−306.537	230.100	11.133	−43.488	10.915
1 2 3 3.724	−825.366	−125.132	−219.439	225.023	1.563	−46.822	11.295
1 3 1 3.612	−916.195	−66.679	−249.594	202.129	−0.587	−34.503	9.452
1 3 2 3.653	−692.693	−21.595	−257.510	243.434	1.660	−41.741	13.837
1 3 3 3.649	−823.232	−61.339	−279.312	207.134	3.000	−35.526	11.297
2 1 1 3.707	−852.892	−234.946	−355.679	179.999	10.863	−41.698	5.165
2 1 2 3.781	−1282.362	−241.629	−214.109	155.075	4.017	−24.010	11.828
2 1 3 3.824	−1029.428	−171.835	−143.172	123.381	5.138	−13.596	8.787
2 2 1 3.677	−859.339	−104.424	−321.580	184.788	9.329	−40.740	6.136
2 2 2 3.742	−814.322	−136.234	−224.007	211.800	0.972	−40.556	10.588
2 2 3 3.720	−803.952	−192.129	−241.804	232.219	3.436	−48.260	12.495
2 3 1 3.592	−820.162	23.941	−189.739	221.675	5.091	−47.157	11.940
2 3 2 3.690	−374.549	−82.929	−197.699	206.384	−1.941	−42.286	11.744
2 3 3 3.693	−838.900	−96.564	−256.843	203.186	5.460	−43.662	10.334

3.8 REGRESSION OF THE TREATMENT EFFECTS ON TIME

3.8.1 Superphosphate

The values of $c_2(P)$ were close to their mean, 0.0364, for most of the period, but the first four values and the penultimate one were considerably lower than the mean. The following seventh degree polynomial was required to fit the values, although the mean was by far the most important coefficient:

$$\hat{c}_2 \times 10^5 = 3644 + 31\xi_1 - 24\xi_2 + 16\xi_3 + 0.77\xi_4 - 1.1\xi_5$$
$$(\pm 533) \quad (\pm 59) \quad (\pm 15) \quad (\pm 16) \quad (\pm 5.5) \quad (\pm 1.1)$$

$$+ 2.5\xi_6 + 1.40\xi_7.$$
$$(\pm 1.3) \quad (\pm 0.61)$$

(3.3)

The coefficients and their standard errors which are given in brackets below them were taken from the rows of estimates for P in Table 3.8.

The mean of the values of $c_{22}(P^2)$, -0.0081 ± 0.0031, was significantly different from zero ($P = 0.03$).

Table 3.8   Analysis of variance of the values of b_0, and $10^5 \times b_1, \ldots, b_7$ of Table 3.7

			b_0	
Source	*DF*	*MS* $\times 10^3$	*F*	*P*(%)
Blocks	1	0.34	0.97	34.9
P	1	15.9	46.7	0.013
P^2	1	2.4	6.9	2.9
S	1	46.6	136.6	0.00
S^2	1	0.033	0.09	76.2
$P \times S$	1	0.34	1.00	34.6
$P \times S^2$	1	0.49	1.4	26.3
$P^2 \times S$	1	0.85	2.5	15.3
$P^2 \times S^2$	1	0.44	1.3	28.8
Error	8	0.34		

Variable	Regn coefficient $\times 10^3$	Standard error $\times 10^3$
Constant term	3709.6	4.35
P	36.4391	5.33
P^2	−8.1309	3.079
S	−62.3399	5.33
S^2	0.9646	3.079
$P \times S$	−6.5367	6.53
$P \times S^2$	4.5371	3.77
$P^2 \times S$	−5.9468	3.77
$P^2 \times S^2$	2.4799	2.18

			$b_3 \times 10^5$	
Source	*DF*	*MS*	*F*	*P*(%)
Blocks	1	1 720	0.54	48.4
P	1	3 188	0.10	34.7
P^2	1	1	0.00	98.8
S	1	187	0.059	81.5
S^2	1	1 157	0.36	56.4
$P \times S$	1	10 211	3.2	11.1
$P \times S^2$	1	548	0.17	68.9
$P^2 \times S$	1	336	0.105	75.4
$P^2 \times S^2$	1	170	0.053	82.3
Error	8	3 191		

$b_1 \times 10^5$			$b_2 \times 10^5$		
MS	F	P(%)	MS	F	P(%)
105 547	2.5	15.4	945	0.33	58.2
11 853	0.28	61.2	6 751	2.3	16.4
78 549	1.8	21.1	578	0.20	66.6
471 848	11.1	1.04	116 373	40.4	0.02
6 206	0.15	71.2	787	0.27	61.5
1 737	0.04	84.5	392	0.14	72.2
18 293	0.43	53.02	89	0.03	86.5
28 272	0.66	43.8	243	0.08	77.9
42 518	1.0	34.6	1	0.00	98.5
42 512			2 878		

Regn coefficient	Standard error	Regn coefficient	Standard error
−929.4	48.60	−144.6633	12.64
31.4286	59.52	−23.7187	15.49
−46.7109	34.36	4.0067	8.94
198.2943	59.52	98.4772	15.49
−13.1295	34.36	−4.6748	8.94
14.7357	72.90	−6.9996	18.97
−27.6081	42.09	1.9272	10.95
−34.3221	42.09	−3.1814	10.95
−24.3007	24.30	−0.1247	6.32

$b_4 \times 10^5$			$b_5 \times 10^5$		
MS	F	P(%)	MS	F	P(%)
1 218.8	3.4	10.2	0.521	0.035	85.7
7.1	0.020	89.1	14.6	0.97	35.3
337.4	0.94	35.9	38.6	2.6	14.7
6 335.1	17.7	0.29	61.7	4.1	7.7
2 397.1	6.7	3.2	0.872	0.058	81.6
92.7	0.259	62.4	25.01	1.7	23.2
1 365.7	3.8	8.6	2.5	0.17	69.4
119.5	0.34	57.9	12.7	0.84	38.5
1.9	0.006	94.2	48.2	3.2	11.1
357.3			14.9		

Table 3.8 (Continued)

Source	DF	MS	$b_3 \times 10^5$ F	P(%)
Variable		Regn coefficient	Standard error	
Constant term		−248.0685	13.31	
P		16.2987	16.31	
P^2		−0.1429	9.41	
S		3.9495	16.31	
S^2		5.6698	9.41	
$P \times S$		−35.7263	19.97	
$P \times S^2$		−4.7780	11.53	
$P^2 \times S$		−3.7423	11.53	
$P^2 \times S^2$		−1.5374	6.66	

Source	DF	MS	$b_6 \times 10^5$ F	P(%)
Blocks	1	8.894	0.42	53.4
P	1	75.846	3.6	9.4
P^2	1	3.400	0.16	69.8
S	1	498.307	23.6	0.12
S^2	1	307.937	14.6	0.51
$P \times S$	1	205.803	9.8	1.4
$P \times S^2$	1	242.074	11.5	0.95
$P^2 \times S$	1	14.649	0.69	42.9
$P^2 \times S^2$	1	1.153	0.055	82.1
Error	8	21.077		
Variable		Regn coefficient	Standard error	
Constant term		−37.2934	1.082	
P		2.5141	1.32	
P^2		−0.3073	0.765	
S		−6.4440	1.32	
S^2		2.9247	0.765	
$P \times S$		−5.0720	1.62	
$P \times S^2$		3.1759	0.937	
$P^2 \times S$		0.7813	0.937	
$P^2 \times S^2$		0.1265	0.541	

$b_4 \times 10^5$			$b_5 \times 10^5$		
MS	F	P(%)	MS	F	P(%)
Regn coefficient	Standard error		Regn coefficient	Standard error	
199.1736	4.46		4.5369	0.913	
0.7687	5.46		−1.1017	1.12	
−3.0612	3.15		1.0351	0.645	
22.9766	5.46		−2.2674	1.12	
−8.1601	3.15		−0.1556	0.645	
3.4036	6.68		1.7683	1.37	
−7.5434	3.86		0.3224	0.7905	
−2.2317	3.86		−0.7265	0.7905	
−0.1665	2.23		0.8186	0.456	

$b_7 \times 10^5$		
MS	F	P(%)
13.169	2.9	12.3
23.513	5.3	5.1
12.338	2.8	13.4
4.556	1.03	33.9
0.186	0.042	84.3
14.719	3.3	10.6
0.006	0.001	97.1
0.635	0.14	71.5
4.699	1.1	33.3
4.425		
Regn coefficient	Standard error	
10.7460	0.496	
1.3998	0.607	
−0.5854	0.351	
0.6162	0.607	
0.0718	0.351	
−1.3564	0.744	
0.0163	0.429	
0.1626	0.429	
−0.2555	0.248	

Table 3.9

Treatment variable	Order of polynomial coefficient							
	0	1	2	3	4	5	6	7
P	***						?	*
P^2	*							
S	***	**	***		***		***	
S^2					*		**	
$P \times S$							**	
$P \times S^2$							**	
$P^2 \times S$								
$P^2 \times S^2$								

?indicates probability values near 0.10.
*indicates probability values near 0.05.
**indicates probability values near 0.01.
***indicates probability values near 0.001.

Although there was no consistent effect of either $P \times S$ or $P \times S^2$, the values of c_{23} and c_{233} showed high-order variation and sixth degree polynomials were needed to fit them. The fitted equations were:

$$\hat{c}_{23} \times 10^5 = \underset{(\pm 653)}{654} + \underset{(\pm 73)}{14\xi_1} - \underset{(\pm 19)}{7\xi_2} - \underset{(\pm 20)}{36\xi_3} + \underset{(\pm 6.7)}{3.4\xi_4}$$
$$+ \underset{(\pm 1.4)}{1.8\xi_5} - \underset{(\pm 1.6)}{5.1\xi_6} \tag{3.4}$$

and

$$\hat{c}_{233} \times 10^5 = \underset{(\pm 377)}{454} - \underset{(\pm 42)}{28\xi_1} + \underset{(\pm 11)}{2\xi_2} - \underset{(\pm 12)}{5\xi_3}$$
$$- \underset{(\pm 3.9)}{7.5\xi_4} + \underset{(\pm 0.79)}{0.32\xi_5} + \underset{(\pm 0.94)}{3.2\xi_6.} \tag{3.5}$$

None of the polynomial coefficients was significant for $c_{223}(P^2 \times S)$ or for $c_{2233}(P^2 \times S^2)$.

3.8.2 Stocking rate

The values of $c_3(S)$ mainly followed a quadratic trend over the period, but the values for weighings 8, 9 and 10 deviated markedly from a quadratic. A polynomial of degree 6 was required to fit the trend. The fitted regression was:

$$\hat{c}_3 \times 10^5 = \underset{(\pm 533)}{-6234} + \underset{(\pm 60)}{198\xi_1} + \underset{(\pm 15)}{98\xi_2} + \underset{(\pm 16)}{3.9\xi_3} + \underset{(\pm 5.5)}{23.0\xi_4} - \underset{(\pm 1.1)}{2.3\xi_5} - \underset{(\pm 1.3)}{6.4\xi_6.} \tag{3.6}$$

There was no consistent pattern in the values of c_{33} (S^2) over the period, but some high-order variation resulted in a significant fourth-order coefficient and a highly significant sixth-order coefficient. The fitted regression was

$$\hat{c}_{33} = \begin{array}{cccccc} 96 & - \; 13\xi_1 & - \; 4.7\xi_2 & + \; 5.7\xi_3 & - \; 8.2\xi_4 \\ (\pm 308) & (\pm 34) & (\pm 8.9) & (\pm 9.4) & (\pm 3.2) \end{array}$$

$$\begin{array}{cc} - \; 0.16\xi_5 & + \; 2.9\xi_6 \;. \\ (\pm 0.65) & (\pm 0.77) \end{array} \tag{3.7}$$

The highly significant sixth- and seventh-order coefficients were unusual, particularly for S^2, $P \times S$ and $P \times S^2$, for which none of the lower-order coefficients was significant. The order was reduced by using periodic terms such as $\cos d$, $\sin d$, $\cos 2d$ and $\sin 2d$ and their interactions with years. However, this led to a lot more computing and was not warranted, since the smoothed values obtained using the periodic terms were very similar to those given by the polynomials.

The high-order variation could be caused by environmental factors not considered here. Methods of investigating this are given in Chapter 8.

3.9. PREDICTED VALUES OF THE TREATMENT CONTRASTS

The fitted regressions which are shown in Figure 3.2 were plotted from the predicted values given in Table 3.10. The predicted values can be obtained directly by substituting the values of $\xi_1, \ldots, \xi_6$ or $\xi_1, \ldots, \xi_7$ from Table 3.6 in Equations (3.4)–(3.7) as was described in section 2.8. Equivalently, Table 3.7 was expanded to include predicted live weight for each weighing using sixth degree and seventh degree polynomials as follows.

Sixth degree polynomial:

$$(12) = (4) - 8 \times (5) + 40 \times (6) - 28 \times (7) + 52 \times (8) - 104 \times (9) + 104(10)$$
$$(13) = (4) - 7(5) + 25(6) - 7(7) - 13(8) + 91(9) - 169(10)$$
$$\vdots$$
$$(28) = (4) - 8(5) + 40(6) + 28(7) + 52(8) + 104(9) + 104(10)$$

Seventh degree polynomial:

$$(29) = (12) - 130(11)$$
$$(30) = (13) - 325(11)$$
$$\vdots$$
$$(45) = (28) + 130(11)$$

The multipliers are the rows of Table 3.6.

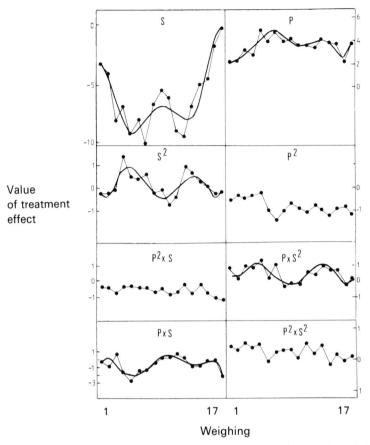

Figure 3.2 Values of the main and interaction effects of Equation (3.1) and polynomial regressions fitted to them.

Analysis of variance of columns 12–28 provided the predicted values and their standard errors for c_{23}, c_{233}, c_3 and c_{33}. Analysis of columns 29–45 provided those for c_2. Columns 12, 13, 28, 29, 30 and 45 are given in Table 3.11 as examples. There was very little difference between the two sets of predicted values here because the polynomials were very similar.

The values of $\hat{c}_{23}$ (P $\times$ S) were mainly negative and those of c_{233}(P $\times$ S²) positive. Hence the regression of the values of the linear effect of P on the levels of stocking rate was quadratic of the form shown in Figure 3.3(a). Therefore, although the relation between live weight and level of P was of the form illustrated in Figure 3.3(b), the effect of P decreased as stocking rate was increased. The values of $\hat{c}_3$(S) were all negative and those of $\hat{c}_{33}$(S²) were mostly positive so that the relation between live weight and stocking rate can be represented by the curve in Figure 3.3(c) over most of the period.

Table 3.10 Predicted values $\times 10^5$ and their standard errors for important regression coefficients of Equation (3.1)

Weighing	c_2	SE	c_{22}	SE	c_3	SE	c_{33}	SE	c_{23}	SE	c_{233}	SE
	(P)		(P^2)		(S)		(S^2)		$(P \times S)$		$(P \times S^2)$	
1	2409.4	563.9	−810	31	−3327.4	573.4	−245.0	331.1	−405.6	702.3	784.0	405.5
2	2187.8	781.5			−4419.6	881.9	343.8	468.7	−145.0	994.3	320.9	574.1
3	2950.4	910.3			−6883.8	916.6	233.6	529.2	−572.9	1122.6	684.7	648.1
4	3804.1	1002.9			−8639.3	1035.4	730.5	597.8	−1325.1	1268.2	1027.4	732.2
5	4350.4	989.0			−9146.4	1003.8	857.1	579.5	−1888.6	1229.4	1032.5	709.8
6	4511.5	881.0			−8688.7	882.0	620.7	509.2	−1927.0	1080.2	715.2	623.6
7	4388.2	776.9			−7855.2	808.8	189.7	467.0	−1403.4	990.6	257.9	571.9
8	4148.8	756.8			−7193.2	795.0	−220.1	459.0	−551.8	973.6	−118.9	562.1
9	3945.3	777.6			−7012.9	778.1	−432.5	449.3	249.1	953.0	−249.0	550.2
10	3862.4	759.9			−7316.8	746.6	−360.8	431.0	647.2	914.4	−77.5	527.9
11	3896.7	694.3			−7828.5	711.6	−68.4	410.9	478.9	871.6	321.4	503.2
12	3963.0	644.2			−8102.1	662.3	312.6	382.4	−126.9	811.1	763.0	468.3
13	3935.7	646.5			−7681.9	593.7	558.1	342.8	−761.4	727.1	1011.7	419.8
14	3715.3	650.3			−6293.6	548.0	476.7	316.4	−913.8	671.1	877.8	387.5
15	3328.4	624.1			−4039.7	607.7	34.6	350.9	−369.6	744.3	354.6	429.7
16	3056.4	642.6			−1576.0	757.2	−466.6	437.2	121.0	927.4	−201.8	535.4
17	3596.0	831.6	−810	31	−245.1	810.0	−160.2	467.7	−2155.4	992.1	146.2	572.8

Table 3.11 Live weights (in logs) for each plot at weighing times 1, 2 and 17 predicted by sixth degree (columns 12, 13, 28) and seventh degree polynomials (29, 30 and 45)

BSP	Column (extension of Table 3.7)					
	(12)	(13)	(28)	(29)	(30)	(45)
1 1 1	3.8787	3.8627	3.5409	3.8687	3.8877	3.5509
1 1 2	3.8678	3.7955	3.5686	3.8514	3.8364	3.5849
1 1 3	3.9272	3.8661	3.5882	3.9075	3.9154	3.6079
1 2 1	3.8388	3.7677	3.5295	3.8229	3.8073	3.5454
1 2 2	3.8941	3.8333	3.5856	3.8799	3.8688	3.5998
1 2 3	3.8681	3.8172	3.6164	3.8534	3.8539	3.6311
1 3 1	3.7983	3.7084	3.5108	3.7861	3.7392	3.5230
1 3 2	3.8533	3.7545	3.6018	3.8353	3.7995	3.6197
1 3 3	3.8362	3.7467	3.5543	3.8215	3.7834	3.5690
2 1 1	3.8198	3.7898	3.5067	3.8131	3.8066	3.5134
2 1 2	3.8984	3.8494	3.5817	3.8830	3.8879	3.5970
2 1 3	3.9224	3.8747	3.6882	3.9110	3.9033	3.6996
2 2 1	3.8380	3.7869	3.5399	3.8301	3.8068	3.5478
2 2 2	3.8823	3.8225	3.6286	3.8686	3.8569	3.6424
2 2 3	3.8422	3.7997	3.5853	3.8259	3.8403	3.6015
2 3 1	3.7812	3.7242	3.5544	3.7657	3.7630	3.5699
2 3 2	3.8075	3.7522	3.6328	3.7922	3.7904	3.6481
2 3 3	3.8480	3.7979	3.5813	3.8345	3.8315	3.5947
$s_t^2 \times 10^4$	3.814	7.333	8.294	3.944	7.896	7.896

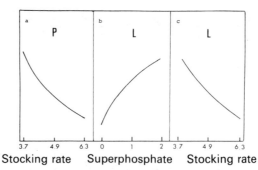

Figure 3.3 Relations between (a) P (linear regression coefficient of live weight) and stocking rate (ewes/ha); (b) L (live weight) and superphosphate in units of 125 kg/ha; and (c) live weight and stocking rate (diagrammatic).

3.10 PREDICTED VALUES FOR LIVE WEIGHT

The preceding analyses have indicated that all variables except $x_2^2 x_3$ ($P^2 \times S$) and $x_2^2 x_3^2$ ($P^2 \times S^2$) were needed to describe the effects of P and S at some stage of the experiment. Mean live weight can be predicted from the following equation:

$$\hat{y} = c_{0,t} + \hat{c}_{2,t} x_2 + \hat{c}_{22} x_2^2 + \hat{c}_{3,t} x_3 + \hat{c}_{33,t} x_3^2 + \hat{c}_{23,t} x_2 x_3 + \hat{c}_{233,t} x_2 x_3^2. \quad (3.8)$$

The form of Equation (3.8) stresses that smoothed values of the coefficients are to be used, and that all the coefficients except c_{22} vary with time.

The effects of the treatments are just as well displayed as deviations from $c_{0,t}$ because $c_{0,t}$ was not affected by the treatments. Hence, for weighings 1, 5 and 10, for example:

$$
\hat{y} - c_{0,t} =
\begin{array}{c}
X \\
\begin{bmatrix}
x_2 & x_3 & x_4 & x_5 & x_6 & x_7 \\
-1 & 1 & -1 & 1 & 1 & -1 \\
0 & -2 & -1 & 1 & 0 & 0 \\
1 & 1 & -1 & 1 & -1 & 1 \\
-1 & 1 & 0 & -2 & 0 & 2 \\
0 & -2 & 0 & -2 & 0 & 0 \\
1 & 1 & 0 & -2 & 0 & -2 \\
-1 & 1 & 1 & 1 & -1 & -1 \\
0 & -2 & 1 & 1 & 0 & 0 \\
1 & 1 & 1 & 1 & 1 & 1
\end{bmatrix}
\end{array}
\times
\qquad (3.9)
$$

$$
\begin{array}{c}
\beta_t \\
\begin{array}{ccc}
t_1 & t_5 & t_{10}
\end{array} \\
\begin{bmatrix}
0.0226 & 0.0501 & 0.0368 \\
-0.0053 & -0.0019 & -0.0087 \\
-0.0328 & -0.0943 & -0.0622 \\
-0.0028 & 0.0051 & -0.0080 \\
-0.0024 & -0.0283 & 0.0027 \\
0.0084 & 0.0136 & -0.0025
\end{bmatrix}
\end{array}
$$

$$
=
\begin{array}{c}
\begin{array}{ccc}
t_1 & t_5 & t_{10}
\end{array} \\
\begin{bmatrix}
-0.0087 & 0.0055 & 0.0139 \\
0.0406 & 0.1032 & 0.0716 \\
0.0581 & 0.1895 & 0.0771 \\
-0.0055 & -0.0350 & -0.0345 \\
0.0162 & -0.0064 & 0.0334 \\
0.0061 & 0.0108 & 0.0491 \\
-0.0695 & -0.1265 & -0.1159 \\
-0.0250 & -0.0930 & -0.0528 \\
-0.0123 & -0.0557 & -0.0419
\end{bmatrix}
\end{array}
\qquad (3.10)
$$

These predicted values are plotted in Figure 3.4 to illustrate the effect of P at each of the levels of S.

3.11 VARIANCES OF THE PREDICTED LIVE WEIGHTS

The values in the columns of (3.10) were calculated as $X\boldsymbol{\beta}_t$ where $\boldsymbol{\beta}_t$ contains the smoothed values of the cs for weighing t. The variances and covariances of these values are $X \operatorname{var}(\boldsymbol{\beta}_t) X^{\mathrm{T}}$. X is given in (3.9) and

$$\operatorname{var}(\boldsymbol{\beta}_t) = s_t^2 \begin{bmatrix} x^{22} \dots x^{27} \\ x^{ij} \\ x^{27} \dots x^{77} \end{bmatrix}$$

where x^{ij} stands for the element corresponding to the variables i and j in $(X^{\mathrm{T}}X)^{-1}$ in the analysis of section 3.3.1; s_t^2 is the variance estimated in the analysis of the live weights predicted for weighing t. Similar expressions apply for the other weighings. These are the only variances and covariances usually required.

From section 3.3.1

$$72(X^{\mathrm{T}}X)^{-1} = \operatorname{diag.}(4, 4, 6, 2, 9, 3, 3, 1).$$

The variances and covariances of the predicted live weights are then:

$$s_t^2/72 \begin{bmatrix} -1 & 1 & -1 & 1 & 1 & -1 \\ 0 & -2 & -1 & 1 & 0 & 0 \\ 1 & 1 & -1 & 1 & -1 & 1 \\ -1 & 1 & 0 & -2 & 0 & 2 \\ 0 & -2 & 0 & -2 & 0 & 0 \\ 1 & 1 & 0 & -2 & 0 & -2 \\ -1 & 1 & 1 & 1 & -1 & -1 \\ 0 & -2 & 1 & 1 & 0 & 0 \\ 1 & 1 & 1 & 1 & 1 & 1 \end{bmatrix} \begin{bmatrix} 6 & 0 & 0 & 0 & 0 & 0 \\ 0 & 2 & 0 & 0 & 0 & 0 \\ 0 & 0 & 9 & 0 & 0 & 0 \\ 0 & 0 & 0 & 3 & 0 & 0 \\ 0 & 0 & 0 & 0 & 3 & 0 \\ 0 & 0 & 0 & 0 & 0 & 1 \end{bmatrix} \times$$

$$\begin{bmatrix} -1 & 0 & 1 & 1 & 0 & 1 & -1 & 0 & 1 \\ 1 & -2 & 1 & 1 & -2 & 1 & 1 & -2 & 1 \\ -1 & -1 & -1 & 0 & 0 & 0 & 1 & 1 & 1 \\ 1 & 1 & 1 & -2 & -2 & -2 & 1 & 1 & 1 \\ 1 & 0 & -1 & 0 & 0 & 0 & -1 & 0 & 1 \\ -1 & 0 & 1 & 2 & 0 & -2 & -1 & 0 & 1 \end{bmatrix}$$

$$= s_t^2/72 \begin{bmatrix} 24 & 8 & 4 & -12 & -10 & -8 & 0 & -10 & -8 \\ 8 & 20 & 8 & -10 & 2 & -10 & -10 & 2 & -10 \\ 4 & 8 & 24 & 4 & -10 & 0 & -8 & -10 & 0 \\ 0 & -10 & -8 & 12 & 8 & 4 & 0 & -10 & -8 \\ -10 & 2 & -10 & 8 & 20 & 8 & -10 & 2 & -10 \\ -8 & -10 & 0 & 16 & 8 & 24 & -8 & -10 & 0 \\ 0 & -10 & -8 & -12 & -10 & -8 & 24 & 8 & 4 \\ -10 & 2 & -10 & -10 & 2 & -10 & 8 & 20 & 8 \\ -8 & -10 & 0 & 4 & -10 & 0 & 4 & 8 & 24 \end{bmatrix}. \quad (3.11)$$

Strictly, $\hat{c}_2$ has a different variance from the other $\hat{c}_s$ because it came from a different analysis, but it was not very much different and was taken to be the same for ease of presentation. Some values of s_t^2 are given at the bottom of Table 3.11.

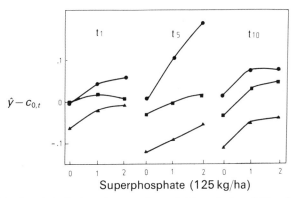

Figure 3.4   Predicted deviations from c_0 for weighings 1, 5 and 10 using the values calculated in expression (3.9) SR 3.7 (●), SR 4.9 (■) and SR 6.2 (▲).

The complete set of variances and covariances is given by the following matrix:

$$\begin{bmatrix} V_1 & \text{cov}_{1,2} & \cdots & \text{cov}_{1,17} \\ \text{cov}_{1,2} & V_2 & & \text{cov}_{2,17} \\ \vdots & & \ddots & \vdots \\ \text{cov}_{1,17} & & & V_{17} \end{bmatrix}$$

in which each V is of the form (3.11) and the cov are similar to (3.11) except that covariances between times replace variances. The covariances can be obtained in a multivariate analysis of variance of the expected live weights for all weighings.

4 Definable within-individual comparisons

4.1 INTRODUCTION

The aims of the experiments considered in Chapters 2 and 3 led to a study of clearly defined treatment comparisons in the variable that had been measured or a simple transformation of it. The sequential analysis was then mainly concerned with approximating trends in the comparisons with time.

Often, however, it is necessary, or more natural, to calculate functions of the sequential observations on an individual as statistics that describe the important features of the individual. Analyses are then carried out to determine how the treatments or, in general, the design variables affect the values of these statistics.

The main features of the analysis are the same as those used in the previous chapters but the interpretation is different.

4.2 AN EXAMPLE

Beattie *et al.* (1979) described an experiment to study the effect of spacing on the yield and size of Jonathan, Delicious and Granny Smith apples on four rootstocks. Trees were planted at spacings ranging from 1.9 to 3.7 m in 1967, in rows 6.1 m apart. There were two rows (replicates) each of Jonathan, Delicious and Granny Smith cultivars. Each row consisted of four panels, one of each of the rootstocks 102, 104, 106, which were new stocks, and Northern Spy. Each panel contained 13 trees. The distance between the trees within each panel increased from 1.9 m at one end to 3.7 m at the other, in steps of 0.15 m. The panels were buffered and the positions of the panels were randomized within the row.

4.2.1 Simple within-individual design

The data for size in 1975 provide an example of a simple within-individual design. Data on size for all years are considered in section 4.2.2 as an example of a complex within-individual design.

The circumference of the butts (girths) for the 13 spacings are given in columns 4 to 16 of Table 4.1.

Table 4.1 Girths (cm) of apple trees at 13 spacings for three cultivars (C) and four rootstocks (S) in two rows (R), and the mean of these and the linear and quadratic coefficients for the regression of girth on spacing

(1)	(2)	(3)	(4)						Spacing*					(16)	(17)	(18)	(19)
R	C	S	1	2	3	4	5	6	7	8	9	10	11	12	13	Mean	Spacing-linear	Spacing-quadratic
1	1	1	28.5	22.9	22.5	23.4	23.9	24.9	23.6	21.7	28.3	15.9	22.4	17.5	27.0	23.27	−0.2929	0.0502
		2	24.6	22.3	19.4	22.9	22.2	15.3	15.0	16.8	26.4	21.1	24.1	20.4	20.4	20.84	−0.0626	0.1066
		3	28.7	25.3	28.7	30.0	28.7	29.8	32.1	30.5	27.3	31.4	30.7	30.2	27.9	29.33	0.1637	−0.0628
		4	27.6	24.3	27.0	31.7	26.4	29.1	30.2	27.4	27.4	31.4	29.6	32.1	30.4	28.82	0.3604	−0.0006
2	1	1	21.6	22.3	23.5	20.4	18.5	17.3	20.4	17.3	20.7	17.9	17.5	20.3	24.7	20.18	−0.1016	0.1250
		2	22.4	20.7	16.2	23.7	24.9	19.2	26.6	17.1	18.4	22.4	22.6	20.5	18.1	20.98	−0.1110	−0.0430
		3	29.1	26.0	34.6	27.5	29.8	29.0	34.1	29.9	26.8	34.3	30.2	29.3	32.4	30.23	0.1868	−0.0136
		4	24.8	28.4	26.6	28.7	30.5	26.7	29.0	29.7	30.3	28.4	28.9	30.8	36.5	29.18	0.5115	0.0390
3	2	1	24.9	27.9	27.0	25.8	27.0	26.3	18.6	24.9	22.6	26.4	22.9	37.5	25.0	25.91	0.1308	0.1169
		2	23.1	19.5	28.0	22.5	18.5	25.0	18.9	24.8	31.4	22.2	28.1	25.6	23.4	23.92	0.3154	−0.0016
		3	33.8	33.3	30.4	36.8	35.2	35.3	34.0	33.5	33.4	36.6	36.7	33.5	36.8	34.56	0.2099	−0.0006
		4	29.1	30.4	28.7	30.5	30.9	33.9	31.3	31.1	32.3	32.5	34.9	30.8	33.3	31.52	0.3187	−0.0285
4	2	1	28.2	32.3	22.8	20.9	27.4	29.6	31.5	26.4	33.0	31.5	27.9	23.2	25.6	27.72	−0.0049	−0.0697
		2	21.7	23.7	24.9	23.0	20.0	23.2	19.6	28.2	27.6	30.8	27.6	23.0	26.7	24.62	0.4445	−0.0020
		3	29.1	29.8	35.3	32.7	33.1	36.7	32.3	33.1	36.5	31.4	32.6	34.6	32.5	33.05	0.1808	−0.0883
		4	34.7	33.7	31.4	32.7	35.6	32.0	35.7	36.6	38.1	36.8	38.6	40.0	39.0	35.76	0.5934	0.0480
5	3	1	38.3	34.3	38.1	38.3	38.4	38.7	39.5	39.7	42.6	37.8	44.3	42.1	41.5	39.51	0.4994	−0.0009
		2	35.8	35.7	30.0	33.1	35.2	34.6	36.5	30.9	37.5	37.4	34.5	39.1	39.4	35.36	0.3868	0.0820
		3	34.2	35.8	35.7	33.2	29.1	35.4	32.9	38.2	39.0	38.3	33.6	36.1	37.0	35.27	0.2626	0.0200
		4	34.5	40.1	34.6	33.4	35.7	35.5	35.0	36.7	35.4	35.6	38.8	35.4	36.4	35.93	0.0654	0.0262
6	3	1	35.8	35.8	38.5	36.3	37.6	37.8	39.8	36.7	40.4	42.8	36.4	44.3	41.0	38.71	0.4907	0.0096
		2	32.7	34.7	36.6	38.8	34.0	33.4	35.8	36.3	36.1	38.8	40.3	40.0	38.7	36.63	0.4637	0.0250
		3	34.4	32.6	33.8	35.0	32.9	36.6	38.9	38.3	35.3	38.7	38.3	37.5	40.2	36.35	0.5214	−0.0061
		4	33.3	33.5	33.7	32.0	34.6	33.5	35.4	30.3	37.7	35.0	34.2	32.5	30.7	33.57	−0.0363	−0.0565

*1.9 m between trees in a row for spacing 1 increasing in steps of 0.15 m to 3.7 m for 13.

4.2.1.1 Derived data

The mean girth over spacing and the linear and quadratic coefficients for the regression of girth on spacing were calculated for each panel of 13 trees. They are given in columns 17, 18 and 19 of Table 4.1. They were calculated as follows:

$$(17) = [(4) + (5) + \cdots + (15) + (16)]/13. \tag{4.1}$$

$$(18) = [-6(4) - 5(5) - 4(6) - 3(7) - 2(8) - 1(9) + 0(10) + 1(11) + 2(12)$$
$$+ 3(13) + 4(14) + 5(15) + 6(16)]/182. \tag{4.2}$$

$$(19) = [22(4) + 11(5) + 2(6) - 5(7) - 10(8) - 13(9) - 14(10) - 13(11)$$
$$- 10(12) - 5(13) + 2(14) + 11(15) + 22(16)]/2002. \tag{4.3}$$

The multipliers used in (4.2) and (4.3) are the linear and quadratic orthogonal polynomial multipliers for 13 equally spaced intervals. The divisors are the sums of squares of the multipliers.

4.2.1.2 Analysis of variance

The analysis of variance used was:

Source of variation	DF	
Cultivars (C)	2	
Rows within cultivars (error a) s_a^2	3	
Stocks (S)	3	(4.4)
C × S	6	
Error b s_b^2	9	
Total	23	

The cultivars are well established in commercial use and the interest was in the stocks and in the interaction of cultivars and stocks.

The analysis of the mean examined the effects of cultivars and stocks on average size. The analyses of the linear and quadratic components of spacing examined the following interactions:

1. cultivars × spacing-linear (C × SP-L);
2. cultivars × spacing-quadratic (C × SP-Q);
3. stocks × spacing-linear (S × SP-L);
4. stocks × spacing-quadratic (S × SP-Q);
5. C × S × SP-L;
6. C × S × SP − Q.

Interactions 1, 3 and 5 tested the homogeneity of the coefficients of the linear regression on spacing for the twelve cultivar–stock classes; 2, 4 and 6 tested homogeneity of the coefficients of the quadratic regression on spacing. These were considered to be the interactions most likely to be important.

No worthwhile conclusions could be drawn about 1 and 2 as the tests of significance were based on s_a^2 which was estimated on only three degrees of freedom.

4.2.1.3 Analysis of variance model

The analysis of variance model fitted was:

$$\hat{y} = c_0 + c_1 x_1 + \cdots + c_{14} x_{14} \qquad (4.5)$$

where $x_1, \ldots, x_{14}$ are variables formed to estimate differences between cultivars, between rows, and between stocks and the interaction of cultivars and stocks. The constant term, c_0, estimates the mean, and $c_1, \ldots, c_{14}$ estimate the differences (effects).

Variables can be constructed in a number of ways to provide the analysis (4.4). Those used here are set out in the design matrix, X, given in Table 4.2. Then

$$(X^T X) = \text{diag.}(24, 16, 48, 8, 8, 8, 12, 32, 72, 8, 24, 48, 24, 72, 144)$$

where the elements are the sums of squares given in Table 4.2.

$$(X^T X)^{-1} = \text{diag.}(1/24, 1/16, \ldots, 1/72, 1/144).$$

$X^T X$ was diagonal because the columns of X were mutually orthogonal.

4.2.1.4 Estimates of the effects of cultivars, stocks and spacing on girth

The analyses of variance of mean girth and of the spacing-linear and spacing-quadratic coefficients obtained by fitting Equation (4.5) to each are given in Table 4.3. These analyses were examined in the order spacing-quadratic, spacing-linear and mean girth.

4.2.1.5 Spacing-quadratic

No source of variation was significant in the analysis of the spacing-quadratic coefficients. The constant term in the analysis of the spacing-quadratic coefficient was also tested for significance.

$$\text{The variance of the constant} = 2651 \times 10^{-6}/24$$
$$= 1.104 \times 10^{-4}.$$
$$\text{The standard error} = \pm 0.0105.$$

Therefore the constant, 0.0114, was about the same size as its standard error and was clearly non-significant. The spacing-quadratic coefficient was not used to approximate the relation between girth and spacing.

Table 4.2 Design matrix X used to fit Equation (4.5) to obtain the analysis of variance (4.4)

	Cultivars			Rows			Stocks			Cultivars × stocks					
R C S	x_0	x_1	x_2	x_3	x_4	x_5	x_6	x_7	x_8	x_9	x_{10}	x_{11}	x_{12}	x_{13}	x_{14}
1 1 1	1	1	1	1	0	0	1	1	1	1	1	1	1	1	1
1 1 2	1	1	1	1	0	0	−1	1	1	−1	1	1	−1	1	1
1 1 3	1	1	1	1	0	0	0	−2	1	0	−2	1	0	−2	1
1 1 4	1	1	1	1	0	0	0	0	−3	0	0	−3	0	0	−3
2 1 1	1	1	1	−1	0	0	1	1	1	1	1	1	1	1	1
2 1 2	1	1	1	−1	0	0	−1	1	1	−1	1	1	−1	1	1
2 1 3	1	1	1	−1	0	0	0	−2	1	0	−2	1	0	−2	1
2 1 4	1	1	1	−1	0	0	0	0	−3	0	0	−3	0	0	−3
3 2 1	1	−1	1	0	1	0	1	1	1	−1	−1	−1	1	1	1
3 2 2	1	−1	1	0	1	0	−1	1	1	1	−1	−1	−1	1	1
3 2 3	1	−1	1	0	1	0	0	−2	1	0	2	−1	0	−2	1
3 2 4	1	−1	1	0	1	0	0	0	−3	0	0	3	0	0	−3
4 2 1	1	−1	1	0	−1	0	1	1	1	−1	−1	−1	1	1	1
4 2 2	1	−1	1	0	−1	0	−1	1	1	1	−1	−1	−1	1	1
4 2 3	1	−1	1	0	−1	0	0	−2	1	0	2	−1	0	−2	1
4 2 4	1	−1	1	0	−1	0	0	0	−3	0	0	3	0	0	−3
5 3 1	1	0	−2	0	0	1	1	1	1	0	0	0	−2	−2	−2
5 3 2	1	0	−2	0	0	1	−1	1	1	0	0	0	2	−2	−2
5 3 3	1	0	−2	0	0	1	0	−2	1	0	0	0	0	4	−2
5 3 4	1	0	−2	0	0	1	0	0	−3	0	0	0	0	0	6
6 3 1	1	0	−2	0	0	−1	1	1	1	0	0	0	−2	−2	−2
6 3 2	1	0	−2	0	0	−1	−1	1	1	0	0	0	2	−2	−2
6 3 3	1	0	−2	0	0	−1	0	−2	1	0	0	0	0	4	−2
6 3 4	1	0	−2	0	0	−1	0	0	−3	0	0	0	0	0	6
Sum of squares	24	16	48	8	8	8	12	36	72	8	24	48	24	72	144

It is unusual to test the significance of the constant as it is usually obviously different from zero. This was not so here, however, because the constant was the mean spacing-quadratic coefficient.

4.2.1.6 Spacing-linear

The interaction cultivars × stocks × spacing-linear was very highly significant. This was shown by the very highly significant F-ratio for C × S in the analysis of the spacing-linear coefficients given in Table 4.3.

Predicted values of the spacing-linear coefficients were calculated for each cultivar–stock combination. These were obtained by multiplying the estimates of $c_0, c_1, c_2, c_6, \ldots, c_{14}$ given in Table 4.3 by X ignoring rows. This multiplication is set out on the following pages.

Table 4.3 Analyses of variance of the mean girth and of the spacing-linear and spacing-quadratic coefficients of Table 4.1

Source of variation	DF	Mean MS	F	P(%)	Spacing-linear $MS \times 10^2$	F	P(%)	Spacing-quadratic $MS \times 10^6$	F	P(%)
Cultivars (C)	2	248.901			13.683			1611		
Error a (s_a^2)	3	1.288			0.868			2651		
Stocks (S)	3	48.524	24.4	0.01	3.586	3.2	7.9	4789	1.1	38.2
C × S	6	24.579	12.4	0.07	14.0487	12.3	0.07	1981	0.5	81.2
Error b (s_b^2)	9	1.988			1.1382			4182		

Coefficient	Estimates of $c_0, c_1, \ldots, c_{14}$ of Equation (4.5) Mean	Spacing-linear (estimates $\times 10^2$)	Spacing-quadratic (estimates $\times 10^3$)
c_0	30.47	22.90	11.4
c_1	−2.14	−9.58	14.2
c_2	−2.97	−5.14	−0.5
c_3	0.21	−3.96	−1.8
c_4	−0.65	−2.99	24.8
c_5	0.10	−2.82	19.4
c_6	1.08	−5.96	5.3
c_7	−1.66	−2.48	19.5
c_8	−0.66	−2.44	2.3
c_9	−0.43	5.16	7.6
c_{10}	−0.03	−5.72	7.1
c_{11}	0.06	−2.86	3.1
c_{12}	−0.24	−4.73	15.0
c_{13}	−1.12	−2.37	6.1
c_{14}	−0.61	−6.51	−3.4

C	S	x_0	x_1	x_2	x_6	x_7	x_8	x_9	x_{10}	x_{11}	x_{12}	x_{13}	x_{14}
1	1	1	1	1	1	1	1	1	1	1	1	1	1
	2	1	1	1	−1	1	1	−1	1	1	−1	1	1
	3	1	1	1	0	−2	1	0	−2	1	0	−2	1
	4	1	1	1	0	0	−3	0	0	−3	0	0	−3
2	1	1	−1	1	1	1	1	−1	−1	−1	1	1	1
	2	1	−1	1	−1	1	1	1	−1	−1	−1	1	1
	3	1	−1	1	0	−2	1	0	2	−1	0	−2	1
	4	1	−1	1	0	0	−3	0	0	3	0	0	−3
3	1	1	0	−2	1	1	1	0	0	0	−2	−2	−2
	2	1	0	−2	−1	1	1	0	0	0	2	−2	−2
	3	1	0	−2	0	−2	1	0	0	0	0	4	−2
	4	1	0	−2	0	0	−3	0	0	0	0	0	6

X

Estimates
(Table 4.3)

Predicted spacing-
linear coefficient

$$
\mathbf{\Gamma} \qquad \mathbf{Y}
$$

$$
\mathbf{X} \times
\begin{bmatrix}
0.2290 \\
-0.0959 \\
-0.0514 \\
-0.0596 \\
-0.0248 \\
-0.0244 \\
0.0516 \\
-0.0572 \\
-0.0286 \\
-0.0473 \\
-0.0237 \\
-0.0651
\end{bmatrix}
=
\begin{bmatrix}
-0.1974 \\
-0.0869 \\
0.1753 \\
0.4361 \\
0.0628 \\
0.3798 \\
0.1953 \\
0.4561 \\
0.4951 \\
0.4253 \\
0.1927 \\
0.0144
\end{bmatrix}
\cdot
\qquad (4.6)
$$

This equation is written $\mathbf{X\Gamma} = \mathbf{Y}$.

The values of $\mathbf{Y}$ are produced in the regression analysis. They are the means of the values for the two rows given in column 18 of Table 4.1 because the design was balanced, e.g.

$$-0.1974 = [-0.2929 + (-0.1016)]/2$$

within rounding error.

The standard error of the difference between estimates of the spacing-linear coefficient for any two stocks and a given cultivar was ± 0.107. This standard error was calculated as follows. The coefficient -0.1974 was calculated as:

$$c_0 + c_1 + c_2 + c_6 + c_7 + c_8 + c_9 + c_{10} + c_{11} + c_{12} + c_{13} + c_{14}$$

from Equation (4.6) and the coefficient -0.0869 was calculated as:

$$c_0 + c_1 + c_2 - c_6 + c_8 - c_9 + c_{10} + c_{12} - c_{12} + c_{13} + c_{14}.$$

The difference

$$-0.1974 - (-0.0869) = 2c_6 + 2c_9 + 2c_{12}$$

so that

$$
\begin{aligned}
\text{variance (difference)} &= 4 \ \text{var}(c_6) + 4 \ \text{var}(c_9) + 4 \ \text{var}(c_{12}) \\
&= 4 \ s_b^2(1/12 + 1/8 + 1/24) \\
&= s_b^2 \\
&= 0.011382.
\end{aligned}
$$

The multipliers $1/12$, $1/8$ and $1/24$ were taken from $(\mathbf{X}^\mathbf{T}\mathbf{X})^{-1}$.
The standard error of the difference was

$$\sqrt{0.011382} = \pm 0.1067.$$

This was the standard error used to test differences between stocks within a cultivar. A difference must exceed

$$0.1067 \times 2.2622 = 0.241$$

to be significant at $P=0.05$. The critical value of t on nine degrees of freedom at $P=0.05$ is 2.2622.

Standard errors are usually produced in a standard regression analysis computer program.

1. **Cultivar 1** – stocks 1 and 2 had the lowest coefficients, -0.1974 and -0.0869. These were not significantly different. Their mean was -0.1422 which was rounded to -0.14. The coefficients for stocks 3 and 4 were 0.18 and 0.14 respectively.
2. **Cultivar 2** – stocks 1 and 3 had the lowest coefficients at 0.0628 and 0.1953 but these were not significantly different. Their mean was 0.13. Stocks 2 and 4 had the largest coefficients 0.3798 and 0.4561, which were also not significantly different. Their mean was 0.42.
3. **Cultivar 3** – stock 1, which had the lowest coefficient for cultivars 1 and 2, had the largest coefficient, 0.4951, but this was not significantly larger than 0.4253, the coefficient for stock 2. Stock 3 had the next largest, 0.19, and stock 4 the smallest, 0.01.

4.2.1.7 Mean girth

The interaction, cultivars × stocks, was very highly significant in the analysis of the mean (Table 4.3) and the main effect of stocks was also very highly significant. Predicted values of the mean girths were calculated for each cultivar–stock combination by multiplying the estimates of c_0, c_1, c_2, $c_6, \ldots, c_{14}$ given in Table 4.3 by X, ignoring rows, exactly as was done for the spacing-linear coefficients in section 4.2.1.6. This multiplication was:

C	S	x_0	x_1	x_2	x_6	x_7	x_8	x_9	x_{10}	x_{11}	x_{12}	x_{13}	x_{14}	Estimates (Table 4.3) Γ	Predicted mean girth Y
1	1	1	1	1	1	1	1	1	1	1	1	1	1	30.47	21.75
	2	1	1	1	-1	1	1	-1	1	1	-1	1	1	-2.14	20.93
	3	1	1	1	0	-2	1	0	-2	1	0	-2	1	-2.97	29.77
	4	1	1	1	0	0	-3	0	0	-3	0	0	-3	1.08	28.99
2	1	1	-1	1	1	1	1	-1	-1	-1	1	1	1	-1.66	26.83
	2	1	-1	1	-1	1	1	1	-1	-1	-1	1	1	-0.66	24.29
	3	1	-1	1	0	-2	1	0	2	-1	0	-2	1	-0.43	33.81
	4	1	-1	1	0	0	-3	0	0	3	0	0	-3	-0.03	33.63
3	1	1	0	-2	1	1	1	0	0	0	-2	-2	-2	0.06	39.11
	2	1	0	-2	-1	1	1	0	0	0	2	-2	-2	-0.24	35.99
	3	1	0	-2	0	-2	1	0	0	0	0	4	-2	-1.12	35.81
	4	1	0	-2	0	0	-3	0	0	0	0	0	6	-0.61	34.73

$$= \qquad (4.7)$$

The standard error of the difference between stocks was ± 1.41 for each cultivar. The mean girths could only be meaningfully compared for those stocks that had similar spacing-linear coefficients. The difference between the stocks in girth was required to exceed

$$1.41 \times 2.2622 = 3.19$$

to be significant at $P = 0.05$.

1. **Cultivar 1** – the mean girths for stocks 1 and 2 were not significantly different. The mean (of the means) was 21.34.
2. **Cultivar 2** – stocks 1 and 3 were significantly different. Stocks 2 and 4 were significantly different.
3. **Cultivar 3** – the difference between stocks 1 and 2 was nearly significant at $P = 0.05$.

4.2.1.8 Regression of girth on spacing

The regressions of girth on spacing for the twelve cultivar–stock classes were obtained from the predicted mean girths of expression (4.7) and the predicted spacing-linear coefficients of expression (4.6). They were:

Cultivar	Stock	Regression equation	
1	1	$\hat{y} = 21.75 - 0.197\xi$	
	2	$\hat{y} = 20.93 - 0.087\xi$	
	3	$\hat{y} = 29.77 + 0.175\xi$	
	4	$\hat{y} = 28.99 + 0.436\xi$	
2	1	$\hat{y} = 26.83 + 0.063\xi$	
	2	$\hat{y} = 24.29 + 0.380\xi$	$\xi = -6, -5, -4, \ldots, 5, 6,$
	3	$\hat{y} = 33.81 + 0.195\xi$	
	4	$\hat{y} = 33.63 + 0.456\xi$	
3	1	$\hat{y} = 39.11 + 0.495\xi$	
	2	$\hat{y} = 35.99 + 0.425\xi$	
	3	$\hat{y} = 35.81 + 0.193\xi$	
	4	$\hat{y} = 34.73 + 0.014\xi.$	

The standard error of each predicted mean was ± 0.952 and the standard error of each predicted spacing-linear coefficient was ± 0.0732. The standard errors were calculated as

$$[(s_a^2 + 3s_b^2)/8]^{1/2}.$$

This followed from the method of calculating the coefficients. For example, in Equation (4.7)

$$21.75 = c_0 + c_1 + c_2 + c_6 + \cdots + c_{14}.$$

Thus variance of $21.75 = \text{var}(c_0) + \text{var}(c_1) + \text{var}(c_2) + \text{var}(c_6) + \cdots + \text{var}(c_{14})$

$$= s_a^2(1/24 + 1/10 + 1/48) + s_b^2(1/12 + \cdots + 1/144)$$
$$= (s_a^2 + 3s_b^2)/8. \tag{4.8}$$

Expression (4.8) takes account of the different variances that were attributed to cultivars which were regarded as whole-plots and to stocks which were sub-plots. If the twelve combinations of cultivars and stocks had been allocated at random there would have only been one error line in the analysis. Expression (4.8) would then have been

$$s^2(1/24 + 1/16 + 1/48 + 1/124 + \cdots + 1/144) = s^2/2.$$

The regression equations were of the form:

$$\hat{y} = \bar{y} + b\xi \tag{4.9}$$

where b is one of the predicted spacing-linear coefficients of Equation (4.6). Then

$$\text{var}(\hat{y}) = \text{var}(\bar{y}) + \xi^2 \text{var}(b) + 2\xi \, \text{cov}(\bar{y}, b)$$

for

$$\xi = -6, \, -5, \ldots, 5, 6.$$

$$\text{var}(\bar{y}) = (1.288 + 3 \times 1.988)/8 \qquad \text{(Table 4.3)}$$
$$= 0.906500,$$
$$\text{var}(b) = (0.00868 + 3 \times 0.011382)/8 \qquad \text{(Table 4.3)}$$
$$= 0.005353,$$
$$\text{cov}(\bar{y}, b) = [\text{cov}(a) + 3 \, \text{cov}(b)]/8$$
$$= (0.022325 + 3 \times 0.056809)/8 \qquad \text{(Table 4.4)}$$
$$= 0.024094.$$

Cov(a) and cov(b) were obtained from the error a and error b rows of the analysis given in Table 4.4. This analysis was obtained by fitting the mean and the spacing-linear coefficients of Table 4.1 to Equation (4.5) in multi-

Table 4.4 Analysis of the covariance of mean girths and the spacing-linear coefficients of Table 4.1

Source of variation	DF	Mean product
Cultivars (C)	2	5.306086
Error a (s_a^2)	3	0.022326
Stocks (S)	2	0.940852
C × S	6	1.281423
Error b (s_b^2)	9	0.056809

variate (bivariate) regression. The multivariate analysis also gave the analyses of variance of the mean and of the spacing-linear coefficient already given in Table 4.3. Var($\hat{y}$) for $\xi = -6$, 0 and 6, for example, were:

$$\xi = -6,$$
$$\mathrm{var}(\hat{y}) = 0.906500 + 36 \times 0.005353 - 12 \times 0.024094$$
$$= 0.810080;$$
$$\xi = 0,$$
$$\mathrm{var}(\hat{y}) = 0.906504;$$
$$\xi = 6,$$
$$\mathrm{var}(\hat{y}) = 0.906500 + 36 \times 0.005353 + 12 \times 0.024094$$
$$= 1.388336.$$

These variances are conveniently calculated using matrices. Equation (4.9) is written

$$Y = x^{\mathrm{T}} \beta$$

where $x^{\mathrm{T}} = [1, \xi]$ and $\beta^{\mathrm{T}} = [\bar{y}, b]$.
Then

$$\mathrm{var}(\hat{Y}) = x^{\mathrm{T}} \, \mathrm{var}(\beta) x.$$

Using S_a and S_b to denote the 2×2 matrices of variances and covariances derived from the error a and error b lines of the multivariate analysis of $\bar{y}$ and b,

$$\mathrm{var}(\beta) = (S_a + 3S_b)/8$$

$$= 1/8 \begin{bmatrix} 1.288333 & 0.022325 \\ 0.022325 & 0.008683 \end{bmatrix} + 3/8 \begin{bmatrix} 1.988000 & 0.056809 \\ 0.056809 & 0.011382 \end{bmatrix}$$

$$= \begin{bmatrix} 0.906542 & 0.024094 \\ 0.024094 & 0.00535352 \end{bmatrix}.$$

These variances were also calculated by regression analysis. The data records given in Table 4.1 were expanded as follows:

$$(20) = (17) \; -6 \; (18)$$
$$(21) = (17) \; -5 \; (18)$$
$$\vdots \qquad \vdots \qquad \vdots$$
$$(31) = (17) \; +5 \; (18)$$
$$(32) = (17) \; +6 \; (18).$$

Columns 20 to 32 then contained the girth for each tree predicted from a linear regression on spacing calculated from its own panel of trees. Analyses of the values of columns 20 to 32 using Equation (4.5) produced the variances of the predicted values. Columns 20, 26 and 32 are given in Table 4.5 as examples, and the analyses of variance of these are given in Table 4.6.

Table 4.5 Girths of trees at spacings 1, 7 and 13 predicted from linear regression calculated for each panel

R C S	Spacing		
	1	*7*	*13*
1 1 1	25.0264	23.2692	21.5121
2	21.2143	20.8384	20.4626
3	28.3483	29.3307	30.3132
4	26.6527	28.8154	30.9780
2 1 1	20.7945	20.1846	19.5747
2	21.6505	20.9846	20.3187
3	26.1077	29.1769	32.2461
4	29.1099	30.2308	31.3516
3 2 1	25.1231	25.9077	26.6923
2	22.0308	23.9231	25.8154
3	33.3022	34.5615	35.8208
4	29.6033	31.5154	33.4274
4 2 1	27.7450	27.7154	27.6857
2	21.9483	24.6154	27.2824
3	32.2011	35.7615	39.3219
4	31.9692	33.0538	34.1384
5 3 1	36.5110	39.5077	42.5044
2	33.0406	35.3615	37.6824
3	33.6934	35.2692	36.8450
4	35.5384	35.9307	36.3230
6 3 1	35.7637	38.7077	41.6516
2	33.7868	33.5692	33.3516
3	33.2176	36.3461	39.4747
4	33.8483	36.6308	39.4132

Table 4.6 Analyses of variance of the predicted girths given in Table 4.5

Source of variation	DF	Mean squares spacing		
		1	*7*	*13*
Cultivars (C)	2	190.154	248.901	217.500
Error $a(s_a^2)$	3	1.332	1.288	1.869
Stocks (S)	3	42.288	48.524	57.342
C × S	6	14.260	24.579	45.013
Error $b(s_b^2)$	9	1.716	1.988	3.079

The var($\hat{y}$) for spacing 1 was

$$(s_a^2 + 3s_b^2)/8 = (1.332 + 3 \times 1.716)/8$$
$$= 0.8100,$$

in agreement with the value calculated above.

Confidence bands for the regression were calculated as $\hat{y} \pm ts(\hat{y})$, but the value of t to be used in the example required some thought because var($\hat{y}$) was a linear function of s_a^2 and s_b^2. These variances were estimated on three and nine degrees of freedom respectively, and their mean values for the 13 spacings were 1.4098 and 2.1474. Following Cochran and Cox (1957), sections 4.14 and 7.16, the value of t to be used was calculated as:

$$t = (s_a^2 \times 3.1825 + 3 \times s_b^2 \times 2.2622)/(s_a^2 + 3s_b^2)$$
$$= (1.4098 \times 3.1825 + 3 \times 2.1474 \times 2.2622)/(1.4098 + 3 \times 2.1474)$$
$$= 2.427.$$

The values 3.1825 and 2.2622 are the critical values of t at $P = 0.05$ for three and nine degrees of freedom.

4.2.1.9 Remarks

It should be emphasized that the regression equations given in section 4.2.1.8 are the same as would be calculated by fitting a linear equation to the data for each cultivar × stock combination. The indirect method was used to obtain appropriate analyses and standard errors.

The analysis illustrated how experimental layout or procedure determines the appropriate analysis. The structure of the variables in this experiment followed a $4 \times 3 \times 13$ factorial in that all combinations of cultivars, stocks and spacings were used. However, the restrictions on the layout of the stocks within cultivars and the systematic arrangement of the spacings made the above analysis appropriate. The main aspects of the factorial were examined but the degrees of freedom for error were very much reduced from those available in a fully randomized experiment.

4.2.2 Complex within-individual design

The measurements of girth for the period 1971 to 1975 will be used as an example of an experiment with a complex within-individual design. There were 65 observations on each panel of 13 trees. The 65 observations came from a 5×13 design of years and spacings. A factorial-type split-up of years and spacings was used to examine the effects of these variables. The linear component of spacing and the linear and quadratic components of years were assumed to be adequate to approximate trends in the girth.

The most important components in the 5×13 design were then:

1. spacing-linear (SP);
2. years (age)-linear (A);
3. years-quadratic (A^2);
4. SP × A;
5. SP × A^2.

Normally, spacing-quadratic would have been included together with its interaction with years, but the analysis of section 4.2.1 indicated that it was not needed.

These components are associated with the following regression equation:

$$\hat{y} = b_0 + b_1 SP + b_2 A + b_{22} A^2 + b_{12} SP \times A + b_{122} SP \times A^2.$$

The analysis may be viewed as fitting this equation to the observations on each of the 12 combinations of cultivars and stocks in the same way that the linear equation was fitted in section 4.2.1.8.

The interaction terms $SP \times A$ and $SP \times A^2$, particularly $SP \times A$, were expected to be important because they measured how the effect of spacing changed with age.

4.2.2.1 Data

The original data will not be given, but the statistics calculated from them are given in Table 4.7. The original data were in columns 4 to 68 of a data record as shown in Table 4.8. The first record contained the measurements taken on the trees of cultivar 1 on stock 1 in row 1 for the years 1971 to 1975; the measurements for the tree at spacing 1 were in columns 4, 17, 30, 43 and 56 and those for the tree-spacing 2 were in columns 5, 18, 31, 44 and 57; and so on.

The mean and the spacing-linear coefficients for each year were calculated from these as shown below and are given in the columns headed (69) to (78) in Table 4.7.

Mean: $1971 - (69) = [(4) + \cdots + (16)]/13,$

$\qquad \vdots \qquad \vdots$

$\qquad 1975 - (73) = [(8) + \cdots + (20)]/13,$

SP: $\quad 1971 - (74) = [-6\,(4) - 5\,(5) + \cdots + 5(19) + 6(20)]/182,$

$\qquad \vdots \qquad \vdots$

$\qquad 1975 - (78) = [-6\,(8) - 5\,(9) + \cdots + 5\,(19) + 6\,(20)]/182.$

The following statistics were calculated from these and are given in the columns headed (79) to (84) in Table 4.7:

Mean: $\quad (79) = [(69) + (70) + (71) + (72) + (73)]/5,$

SP: $\qquad (80) = [(74) + (75) + (76) + (77) + (78)]/5,$

A: $\qquad (81) = [-2(69) - (71) + 0(73) + (75) + 2(77)]/10,$

A^2: $\qquad (82) = [2(69) - (71) - 2(73) - (75) + 2(77)]/14,$

$SP \times A$: $\quad (83) = [-2(74) - (75) + 0(76) + (77) + 2(78)]/10,$

$SP \times A^2$: $\quad (84) = [2(74) - (75) - 2(76) - (77) + 2(78)]/14.$

Table 4.7 Statistics calculated from the girths of trees at 13 spacings in 5 years for 2 rows (R) of each of 3 cultivars (C) on 4 rootstocks (S)

R C S	(69)	(70)	(71) Mean	(72)	(73)	(74)	(75)	(76) SP	(77)	(78)	(79) Mean	(80) SP	(81) A	(82) A²	(83) SP ×A	(84) SP ×A²
	1971	·	·	·	1975	1971	·	·	·	1975						
1 1 1	13.323	15.628	18.331	20.469	23.269	-0.067	-0.148	-0.271	-0.314	-0.293	18.203	-0.219	2.474	0.031	-0.062	0.020
2	13.115	15.138	16.931	18.915	20.838	-0.089	-0.104	-0.088	-0.046	-0.063	16.988	-0.078	1.922	-0.001	0.011	0.002
3	17.115	20.415	23.462	26.054	29.331	0.050	0.087	0.091	0.124	0.164	23.275	0.103	3.007	-0.036	0.026	0.003
4	17.254	19.215	23.162	24.423	28.815	0.076	0.101	0.166	0.207	0.360	22.574	0.182	2.833	0.155	0.067	0.017
2 1 1	12.392	14.677	16.577	18.023	20.185	-0.001	-0.007	-0.032	-0.068	-0.102	16.371	-0.042	1.893	-0.050	-0.026	-0.005
2	12.046	14.000	16.854	18.300	20.985	-0.020	0.020	-0.109	-0.069	-0.111	16.437	-0.058	2.218	0.004	-0.027	0.000
3	16.354	20.569	23.769	26.700	29.177	0.022	0.176	0.335	0.355	0.512	23.314	0.280	3.178	-0.268	0.116	-0.010
4	17.115	20.023	23.969	25.785	30.231	-0.045	0.027	0.118	0.137	0.187	23.425	0.085	3.199	0.068	0.057	-0.008
3 2 1	16.931	18.969	22.054	22.946	25.908	-0.003	0.054	0.063	0.040	0.131	21.362	0.057	2.193	-0.025	0.025	0.003
2	16.623	18.438	20.477	21.992	23.923	0.107	0.128	0.186	0.200	0.315	20.291	0.187	1.815	-0.021	0.049	0.010
3	21.246	25.477	29.123	31.815	34.562	0.042	0.136	0.180	0.265	0.210	28.445	0.167	3.297	-0.280	0.047	-0.018
4	19.285	22.915	26.169	28.754	31.515	0.040	0.127	0.241	0.259	0.319	25.728	0.197	3.030	-0.172	0.069	-0.011
4 2 1	17.654	19.885	23.231	24.885	27.715	-0.016	-0.018	0.063	-0.014	-0.005	22.674	0.002	2.512	-0.035	0.003	-0.010
2	16.738	18.469	21.123	22.446	24.615	0.071	0.162	0.276	0.324	0.445	20.678	0.255	1.973	-0.032	0.091	0.000
3	22.146	26.331	30.108	32.523	35.762	0.092	0.231	0.373	0.486	0.593	29.374	0.355	3.342	-0.232	0.126	-0.007
4	20.462	23.315	27.000	28.831	33.054	-0.006	0.077	0.099	0.116	0.181	26.532	0.093	3.070	0.063	0.041	-0.003
5 3 1	23.108	28.185	32.885	35.077	39.508	0.026	0.226	0.316	0.400	0.499	31.752	0.294	3.969	-0.271	0.112	-0.015
2	18.769	23.346	28.269	30.046	35.362	0.041	0.221	0.247	0.279	0.387	27.158	0.235	3.988	-0.119	0.075	-0.010
3	21.385	25.954	29.731	31.608	35.269	0.059	0.071	0.147	0.153	0.263	28.789	0.138	3.342	-0.265	0.049	0.009
4	21.269	26.162	29.954	31.831	35.931	-0.067	0.031	-0.005	-0.010	0.065	29.029	0.003	3.499	-0.250	0.022	-0.001
6 3 1	22.385	28.200	32.908	34.692	38.708	0.089	0.281	0.403	0.425	0.491	31.378	0.338	3.914	-0.466	0.095	-0.025
2	17.869	22.115	27.069	28.692	33.569	-0.101	-0.091	-0.032	-0.046	-0.036	25.836	-0.061	3.798	-0.148	0.017	-0.005
3	21.100	25.846	29.977	31.900	36.346	0.116	0.295	0.400	0.473	0.521	29.034	0.361	3.655	-0.201	0.099	-0.021
4	20.277	25.808	30.038	31.877	36.631	0.100	0.170	0.329	0.346	0.464	28.926	0.282	3.878	-0.282	0.090	-0.003

Table 4.8

			1971		1972		1973		1974		1975	
					Spacing							
			1...13		*1...13*		*1...13*		*1...13*		*1...13*	
(1)	*(2)*	*(3)*	*(4)*	*(16)*	*(17)*	*(29)*	*(30)*	*(42)*	*(43)*	*(55)*	*(56)*	*(68)*
R	*C*	*S*					*Girth*					
1	1	1	14.1	14.3	18.4	17.7	22.3	20.1	25.4	23.2	28.5	27.0
...	...	...	...	...	...	...	...	...	...	...	...	...
...	...	...	...	...	...	...	...	...	...	...	...	...
...	...	...	...	...	...	...	...	...	...	...	...	...
6	3	4	19.5	20.8	24.8	26.3	28.0	31.0	29.2	32.5	32.7	38.7

4.2.2.2 Analyses of variance

The analyses of variance of mean girth, SP, A, A^2, SP × A and SP × A^2
were obtained by fitting Equation (4.5) to the values of each of these given
in Table 4.7. The analyses of variance are given in Table 4.9.

1. **SP × A^2** – no source of variation was significant in the analyses of the
   values of SP × A^2. This coefficient was not considered
   further.

2. **SP × A** – the source of variation for cultivars was significant at $P = 0.05$
   and the interaction of cultivars and stocks was highly significant, $P = 0.0036$. Predicted values of this coefficient were
   calculated for each cultivar on each stock. The calculation is
   set out at the end of this section.

3. **A^2** – cultivars was the only significant source of variation. The
   predicted values of the A^2 coefficients for each cultivar were
   obtained from the following multiplication:

$$
\begin{array}{ccccc}
C & x_0 & x_1 & x_2 & \text{Estimates} \\
 & & & & \text{(Table 4.9)}
\end{array}
\qquad
\begin{array}{c}
\text{Predicted } A^2 \\
\text{coefficients}
\end{array}
$$

$$
\begin{array}{c}
1 \\ 2 \\ 3
\end{array}
\begin{bmatrix}
1 & 1 & 1 \\
1 & -1 & 1 \\
1 & 0 & -2
\end{bmatrix}
\begin{bmatrix}
-0.1180 \\
0.0400 \\
0.0660
\end{bmatrix}
=
\begin{bmatrix}
-0.0120 \\
-0.0920 \\
-0.2500
\end{bmatrix}. \tag{4.10}
$$

The 3×3 matrix was obtained from the 12×12 matrix X of Equations
(4.6) and (4.7) by omitting variables x_6–x_{14} and rows 2–4, 6–8 and 10–12.
The variables x_6–x_{14} were associated with S and C × S which were
non-significant. Here, these predicted values were simply the means
calculated over rows and stocks.

The negative coefficients for A^2 indicated that the rate of increase of
girth with spacing was decreasing with age.

Table 4.9 Analyses of variance of the values of mean girth, SP, A, A^2, SP × A and SP × A^2, given in Table 4.7

Source of variation	DF	(a) Mean MS	F	P (%)	(b) SP MS × 10³	F	P (%)	(c) A MS × 10²	F	P (%)	(d) A² MS × 10³	F	P (%)	(e) SP × A MS × 10⁴	F	P (%)	(f) SP × A² MS × 10⁵	F	P (%)
Cultivars (C)	2	159.1	223	<0.01	62.2	10.1	5	343.2	143	<0.01	117.7	10.8	5	52.4	9.9	5.36	26.1	1.1	42
Error a	3	0.682			6.181			2.405			10.88			5.30			22.8		
Stocks (S)	2	29.8	16.6	0.05	10.2	1.8	21.3	62.5	12.3	0.16	6.86	<1		20.9	3.2	7.2	4.7	<1	
C × S	6	14.6	8.1	0.32	57.2	10.3	0.14	43.7	8.6	0.26	10.86	<1		49.9	7.9	0.36	10.3	1.1	
Error b	9	1.795			5.565			5.076			18.43			6.35			9.18		

Estimates of $c_0, c_1, \ldots, c_{14}$ of Equation (4.5)

Coefficient	Mean	SP × 10²	A × 10	A² × 10²	SP × A × 10³	SP × A² × 10³
c_0	24.48	13.2	−16.2	−11.8	−11.8	−3.7
c_1	−2.16	−6.6	−0.3	4.0	−17.9	3.4
c_2	−2.25	−3.4	−37.8	6.6	−10.5	2.6
c_3	0.19	−3.5	−3.1	5.0	−9.6	7.9
c_4	−0.43	−1.2	−7.0	−3.3	−8.9	0.4
c_5	0.19	−3.1	−5.6	2.4	−5.4	4.7
c_6	0.94	−3.3	9.7	−3.0	−11.9	−2.5
c_7	−1.30	−1.8	−17.8	0.1	−5.6	1.2
c_8	−0.50	−0.9	−9.3	1.1	−6.9	0.3
c_9	−0.24	3.2	−8.6	−0.2	5.0	3.8
c_{10}	−0.02	−3.1	1.4	−1.7	−11.0	−0.7
c_{11}	0.05	−1.6	2.0	−1.1	−5.0	−0.9
c_{12}	−0.41	−3.1	4.6	2.7	−11.1	2.1
c_{13}	−0.80	−1.5	−16.2	0.9	−6.0	1.9
c_{14}	−0.51	−4.3	−6.4	1.4	−11.8	0.8

4. **A and SP** – predicted values of both of these coefficients were calculated for each cultivar on each stock because the significant SP × A interaction implied effects of SP or A or both. This conclusion was supported by the analyses of SP and A which indicated significant effects of both. The estimates are calculated at the end of this section.
5. **Mean girth** – C × S was very highly significant and the main effect of stocks was also very highly significant. Predicted values were calculated for each cultivar on each stock.

The predicted values of the mean and the coefficients of SP, A and SP × A for each cultivar on each stock were obtained by multiplying the estimates for each of these given in Table 4.9 by X of Equations (4.6) and (4.7) as shown below and in Table 4.10. These predicted values are equal to the means over rows of the values in Table 4.7.

$$X$$

x_0	x_1	x_2	x_6	x_7	x_8	x_9	x_{10}	x_{11}	x_{12}	x_{13}	x_{14}
1	1	1	1	1	1	1	1	1	1	1	1
1	1	1	−1	1	1	−1	1	1	−1	1	1
1	1	1	0	−2	1	0	−2	1	0	−2	1
1	1	1	0	0	−3	0	0	−3	0	0	−3
1	−1	1	1	1	1	−1	−1	−1	1	1	1
1	−1	1	−1	1	1	1	−1	−1	−1	1	1
1	−1	1	0	−2	1	0	2	−1	0	−2	1
1	−1	1	0	0	−3	0	0	3	0	0	−3
1	0	−2	1	1	1	0	0	0	−2	−2	−2
1	0	−2	−1	1	1	0	0	0	2	−2	−2
1	0	−2	0	−2	1	0	0	0	0	4	−2
1	0	−2	0	0	−3	0	0	0	0	0	6

Estimates

Mean	SP	A	SP × A
24.483	0.1312	3.0000	0.0488
−2.156	−0.0663	−0.0318	−0.0179
−2.254	−0.0336	−0.3777	−0.0105
0.938	−0.0328	−0.0967	−0.0119
−1.300	−0.0178	−0.1775	−0.0056

$$\begin{bmatrix} -0.499 & -0.0093 & -0.0933 & -0.0069 \\ -0.240 & 0.0323 & -0.0862 & 0.0050 \\ -0.019 & -0.0314 & 0.0139 & -0.0110 \\ 0.049 & -0.0146 & 0.0195 & -0.0050 \\ -0.412 & -0.0307 & 0.0462 & -0.0111 \\ -0.799 & -0.0151 & -0.1619 & -0.0060 \\ -0.507 & -0.0426 & -0.0645 & -0.0118 \end{bmatrix} \cdot$$

Table 4.10

C	S	Mean	SP	A	SP × A
1	1	17.29	−0.130	2.18	−0.044
	2	16.71	−0.068	2.07	−0.008
	3	23.35	0.094	3.10	0.042
	4	22.94	0.231	3.00	0.092
2	1	22.02	0.030	2.35	0.014
	2	20.48	0.221	1.89	0.070
	3	27.49	0.130	3.18	0.044
	4	27.55	0.276	3.19	0.097
3	1	31.57	0.316	3.94	0.103
	2	28.04	0.258	3.93	0.083
	2	28.91	0.250	3.50	0.074
Standard error		±0.871	±0.0535	±0.148	±0.174

As explained in section 4.2.1.8, the variances of the $\hat{y}$ were

$$(s_a^2 + 3s_b^2)/8.$$

For example, the variance of the predicted values for the mean was

$$(0.682 + 3 \times 1.795)/8 = 0.7584.$$

The standard error was thus ± 0.871.

Predicted values and their variances or standard errors are usually calculated by standard regression analysis programs.

4.2.2.3 Regression of girth on age and spacing

The analyses of the previous section indicated that the relation between girth and spacing and age could be described by the following regression equation:

$$\hat{y} = b_0 + b_1 SP + b_2 A + b_{22} A^2 + b_{12} SP \times A, \qquad (4.11)$$

with the coefficients taking the values given in Table 4.11. The variables take the following coded values:

$$\begin{aligned} &\text{SP: } -6, \ -5, \ -4, \ -3, \ -2, \ -1, \ 0, \ 1, \ 2, \ 3, \ 4, \ 5, \ 6. \\ &\text{A: } -2, \ -1, \ 0, \ 1, \ 2. \\ &\text{A}^2\text{: } 2, \ -1, \ -2, \ -1, \ 2. \end{aligned}$$

Equation (4.11) can be conveniently written as

$$Y = x^T \boldsymbol{\beta}$$

where

$$x^T = [1, \ SP, \ A, \ A^2, \ SP \times A]$$

and

$$\boldsymbol{\beta}^T = [b_0, b_1, b_2, b_{22}, b_{12}],$$

(cf. section 4.2.1.8).
  Then

$$\text{var}(Y) = x^T \text{var}(\boldsymbol{\beta}) x,$$

where

$$\text{var } \boldsymbol{\beta} = (S_a + 3S_b)/8$$

Table 4.11

C	S	Coefficient				
		b_0	b_1	b_2	b_{22}	b_{12}
1	1	17.29	−0.130	2.18	−0.0120	−0.044
	2	16.71	−0.068	2.07	−0.0120	−0.008
	3	23.35	0.094	3.10	−0.0120	0.042
	4	22.94	0.231	3.00	−0.0120	0.092
2	1	22.02	0.030	2.35	−0.0918	0.014
	2	20.48	0.221	1.89	−0.0918	0.070
	3	27.49	0.130	3.18	−0.0918	0.044
	4	27.55	0.276	3.19	−0.0918	0.097
3	1	31.57	0.316	3.94	0.0142	0.103
	2	28.04	0.258	3.93	0.0142	0.083
	3	29.91	0.250	3.50	0.0142	0.074
	4	27.45	−0.029	3.65	0.0142	0.020

$$
=1/8 \begin{bmatrix}
0.6815 \\
-1.9147 \times 10^{-2} & 6.1813 \times 10^{-3} \\
3.6493 \times 10^{-2} & 9.8113 \times 10^{-3} & 2.4048 \times 10^{-2} \\
7.4100 \times 10^{-2} & -5.4810 \times 10^{-3} & -1.5591 \times 10^{-3} & 1.0883 \times 10^{-2} \\
2.6647 \times 10^{-3} & 1.6163 \times 10^{-3} & 3.2598 \times 10^{-3} & -8.3040 \times 10^{-4} & 5.3043 \times 10^{-4}
\end{bmatrix}
$$

$$
+3/8 \begin{bmatrix}
1.7952 \\
4.6810 \times 10^{-2} & 5.5647 \times 10^{-1} \\
1.0974 \times 10^{-1} & -1.3038 \times 10^{-3} & 5.0764 \times 10^{-2} \\
-1.1191 \times 10^{-1} & -3.7670 \times 10^{-3} & -5.0439 \times 10^{-3} & 1.8426 \times 10^{-2} \\
1.3658 \times 10^{-2} & 1.3791 \times 10^{-3} & 1.5824 \times 10^{-4} & -9.5418 \times 10^{-4} & 6.3543 \times 10^{-3}
\end{bmatrix}
$$

S_a and S_b were obtained from the multivariate analysis of b_0, b_1, b_2, b_{22} and b_{12}.

The variances can also be calculated by expanding the data records given in Table 4.7 and analysing the new columns as was done in section 4.2.1.8. Only selected predicted values are likely to be needed. For example, the predicted values of girth in 1975 at spacing 13 can be calculated as:

$$\hat{y}(SP=6, A=2, A^2=2)=(79)+6(80)+2(81)+2(82)+12(83).$$

Analysis of these values provides the correct variances.

5 Covariance

5.1 INTRODUCTION

Experiments in which measurements have been made on the experimental units before the treatment were assigned will be dealt with in this chapter and the next. The analysis of sequences of observations with initial observations as covariates will be covered in this chapter; the use of initial information in the design of experiments will be considered in Chapter 6.

In agricultural research, the covariate is generally the initial observation on the variable being studied. But this is not necessarily so for example, a measure of vigour of fruit trees such as height or girth after the tree has been established is often a useful covariate for subsequent yield.

Covariance is used in well-designed experiments to reduce the residual variation by allowing for differences that were not controlled by the design. For example, the live weights of animals allotted to different treatments will not be identical. Variation in subsequent live weight between replicates may be reduced by using initial live weight as a covariate. In an ideal experiment in which average live weights of the animals allotted to all treatments are similar, adjustments to the treatment means are small. However, the averages usually vary somewhat and covariance analysis then has the added advantage of allowing the effects of the treatments to be compared at the same value of the initial observation.

Few covariates are generally needed in agricultural experiments because the experimental material can usually be selected for uniformity in most of the variables likely to affect the outcome of the experiment. Genetically similar animals of the same age should usually be used in experimental work. Trees in an experiment would be propagated with material from the one source or, if different sources were used, allowance would be made for this in the design. Trees of similar size and age would be selected. However, natural variation often persists, and methods are needed to allow for this in the analysis and interpretation of experiments.

Any variable that is not affected by the treatments can be used as a covariate in the standard way. When the treatments do affect the values of the covariate, the covariate is regarded as an explanatory variable (Williams, 1959, Chapter 7) and the interpretation is different.

The experiments used as examples in Chapters 2 and 3 had been running for six or eight months before the first measurements were made. However, in the experiment to be considered in this chapter, the plots were first

measured before the treatments could have had any effect. This introduced two new aspects to the analysis.
1. Initial measurements could be used as covariates for subsequent ones.
2. Trends in time in the treatment effects were fitted through the origin.

5.2 BACKGROUND TO THE EXAMPLE

Lodge and Roberts (1979) reported an experiment designed to examine the dry-matter production of natural pasture at five levels of superphosphate P, gypsum S and stocking rate SR. The experiment was conducted near Tamworth in North Western New South Wales. The design was a 20-point central composite, approximately rotatable, in three dimensions (Cochran and Cox, 1957, p. 350).

The P and S treatments were applied in September and stocking was commenced in October 1971. At 38 six-weekly intervals, from early November 1971 to May 1976, pasture was harvested with shears. The harvested material was weighed fresh and subsampled to estimate dry matter yield. The data are given in Table 5.1. The dry matter in November 1971 was taken as the initial dry matter and used as a covariate in the analysis of all other harvests. All dry matter yields were transformed to $\log_e$ (yield) to stabilize the variance.

5.3 ANALYSIS OF VARIANCE OF THE BASIC DESIGN

The design was chosen to fit the following second-order response surface equation to the data:

$$\hat{y} = c_0 + c_1 P + c_2 S + C_3 SR + c_{11} P^2 + c_{22} S^2 + c_{33} (SR)^2 + c_{12} P \times S$$
$$+ c_{13} P \times SR + c_{23} S \times SR. \tag{5.1}$$

The levels of P, S and SR used and coded levels of these are given in Table 5.2 and the matrix of x-variables, X, is given in Table 5.3. X is in a slightly different form from those given in the earlier chapters in that the values of x_1^2, x_2^2 or x_3^2 do not add to zero. They could be made to do so by subtracting the mean of each column from each entry in it. However, the form given is the one usually used for this design and it is convenient because the entries are mainly ones and zeros.

The variables x_1, x_2, x_3 and their interactions are unconfounded but x_1^2, x_2^2, and x_3^2 are slightly confounded. The squared terms are also confounded with the constant but this is unimportant and is a consequence of the coding used.

Table 5.1 Dry matter yield in kg/ha of native pasture at five levels of P, S and SR initially (November 1971) and at 37 subsequent harvests

	(1)	(2)	(3)	(4)	(5)	(6)	(7)	(8)	(9)	(10)	(11)	(12)	(13)	(14)	(15)	(16)	(17)	(18)	(19)	(20)	(21)	(22)
	P	S	SR										Harvest									
Plot	(kg)	(kg)	(sheep)	Nov 71	1	2	3	4	5	6	7	8	9	10	11	12	13	14	15	16	17	18
1	6	14	3.2	2646	2324	2008	2042	2933	2037	1460	1583	1068	1809	3148	2903	2847	2573	2155	2633	4059	3470	2632
2	23	14	3.2	1977	938	1119	644	872	535	1225	2014	1136	2318	3083	2548	1850	1373	2115	3329	4491	4354	4204
3	6	54	3.2	3267	3156	2666	3041	1799	1886	997	2043	1806	1117	2398	1939	1375	2501	1154	2626	3575	4225	2491
4	6	14	6.3	3304	3241	2348	2185	2298	2111	833	1417	390	1129	1667	1694	2050	1196	729	2021	2180	3502	2513
5	23	54	3.2	1709	2916	3479	2517	1775	1381	1249	473	920	1738	2296	3431	1451	1307	2194	1929	4643	3005	3431
6	23	14	6.3	906	888	1879	2804	450	516	384	655	99	617	1133	401	190	419	422	1660	1583	1766	1587
7	6	54	6.3	1758	3349	1794	1494	1479	1735	883	1907	483	1827	1754	1552	510	561	1380	1528	2696	3032	1909
8	23	54	6.3	1632	1459	1927	1993	777	1542	1029	888	299	581	1820	1283	584	415	712	1341	2903	2626	2852
9	29	34	4.8	4363	4312	4130	3456	3112	2335	1184	1248	1461	717	2274	2226	1301	1948	1441	1945	2863	2815	2340
10	0	34	4.8	3755	2807	2787	3215	2362	1909	946	1628	876	693	1734	2069	1284	918	2237	1324	3302	3045	2310
11	15	67	4.8	2575	4183	2612	2178	1970	1899	1812	929	999	1190	1494	1957	2002	1533	4006	2619	3876	2838	3543
12	15	0	4.8	1213	1463	2039	1703	1753	1785	994	679	629	734	1923	1448	1230	1311	991	2750	2870	1612	1547
13	15	34	7.3	1883	2956	3453	1318	2373	1316	1013	884	275	591	1289	1123	1415	882	595	1233	2483	1984	1626
14	15	34	2.2	1741	2907	2436	2117	1120	1883	1523	2950	1299	1391	2325	2677	1738	3111	2714	3020	4534	4592	3045
15	15	34	2.2	2775	4595	4576	4223	2087	3041	2219	1460	1342	1390	1685	2489	1549	2038	1359	3515	4226	4045	2728
16	15	34	2.2	1162	2243	1171	2670	747	887	410	532	287	730	2413	2485	318	671	431	1200	2425	2665	2050
17	15	34	2.2	1571	2062	2779	1774	2968	1217	1116	544	617	804	2395	2489	1334	1141	1260	1707	3030	2736	2216
18	15	34	2.2	2922	3385	1686	2033	1650	1036	644	1210	336	576	1870	1900	1467	1543	1293	1740	4128	4288	2373
19	15	34	2.2	2372	3239	2961	1985	2233	1880	872	1142	876	595	2576	2076	2564	1969	1807	2556	4072	2628	2627
20	15	34	2.2	1788	1688	2903	2980	1746	1535	201	440	492	529	1400	2273	1232	1710	1593	1864	2741	2103	2976

Table 5.1 cont.

Plot	(1) P (kg)	(2) S (kg)	(3) SR (sheep)	(23)	(24)	(25)	(26)	(27)	(28)	(29)	(30)	(31)	(32)	(33)	(34)	(35)	(36)	(37)	(38)	(39)	(40)	(41)
														Column								
														Harvest								
				19	20	21	22	23	24	25	26	27	28	29	30	31	32	33	34	35	36	37
1	6	14	3.2	3535	2624	3759	2195	2107	2853	3005	1619	2644	3377	2485	4052	2239	2781	2727	3111	3243	2101	1597
2	23	14	3.2	4470	2762	2847	3574	3327	2967	2884	2880	3521	3221	4636	3776	2631	3612	4335	4117	2027	3387	2323
3	6	54	3.2	2420	2560	1920	2418	2814	2191	2099	2077	1328	3259	2495	2940	2892	2844	2712	2506	4087	2649	2758
4	6	14	6.3	3249	1978	1896	1233	2559	1782	1190	1461	1405	2137	1790	2759	2719	2906	1951	1133	2488	2111	1916
5	23	54	3.2	2693	3524	2647	2577	2830	1903	2290	2138	2314	2495	2545	3463	2569	3340	3840	3014	2648	2523	3136
6	23	14	6.3	1391	656	1083	298	1154	1576	707	361	315	889	159	1251	816	3663	1149	348	847	876	581
7	6	54	6.3	3412	2270	2108	765	1916	2053	1460	1082	1360	3117	1265	1893	1661	2301	1999	1741	1473	1257	755
8	23	54	6.3	4560	1312	1652	1388	2969	2088	1552	1723	1050	1458	1753	1626	1994	3102	2623	1833	1382	1653	1452
9	29	34	4.8	2152	1893	1895	2158	2117	1828	1709	985	2031	2540	2257	2461	2642	2775	2828	1924	2421	3243	2265
10	0	34	4.8	2351	2451	2268	1579	2319	2247	1904	1183	1446	3078	2300	2271	2499	2405	2107	1747	2430	2720	2294
11	15	67	4.8	4252	2952	2384	2815	2245	2053	1883	1752	2333	3744	2108	2905	2558	3368	3205	3198	2058	2306	1887
12	15	0	4.8	2001	1285	1258	1871	2244	1524	1220	729	2082	2154	908	1998	989	1853	1341	1216	1324	2275	1306
13	15	34	7.3	1730	1322	688	974	1143	1790	857	488	876	1516	1350	1666	1534	2441	1528	1574	1665	1837	1655
14	15	34	2.2	4459	2870	2712	4085	3260	2535	3406	3598	3260	2831	3535	4469	2583	3203	3365	2893	2435	3189	2991
15	15	34	2.2	3488	3747	2312	1389	2796	2141	2510	2145	2338	2032	2135	2060	2580	2244	3289	2433	2172	2220	2061
16	15	34	2.2	2215	1435	762	1376	874	1819	1678	1052	882	1670	1466	2234	1886	3602	2660	2593	2240	1548	929
17	15	34	2.2	2878	3040	1780	1773	2181	1915	2040	1205	1498	2518	1472	2206	1997	2132	2086	1191	2288	2784	1190
18	15	34	2.2	4600	2167	1895	1830	2907	2460	1930	2360	2923	4134	1991	2158	2435	4701	2961	3227	2245	2463	2787
19	15	34	2.2	2176	2437	2129	1110	1566	1809	2754	1067	1989	2608	1962	2078	1646	2818	2006	2058	1999	2703	2149
20	15	34	2.2	2306	1569	1612	940	1060	2014	1826	939	1724	2608	875	1830	1714	3012	2293	1880	1529	819	659

Table 5.2 Actual and coded levels of P, S and SR for the experiment of Lodge and Roberts (1979)

Treatment Levels			Coded Levels		
P (kg/ha)	S (kg/ha)	SR (sheep/ha)	P x_1	S x_2	SR x_3
6	14	3.2	-1	-1	-1
23	14	3.2	1	-1	-1
6	54	3.2	-1	1	-1
6	14	6.3	-1	-1	1
23	54	3.2	1	1	-1
23	14	6.3	1	-1	1
6	54	6.3	-1	1	1
23	54	6.3	1	1	1
29	34	4.8	1.682	0	0
0	34	4.8	-1.682	0	0
15	67	4.8	0	1.682	0
15	0	4.8	0	-1.682	0
15	34	7.3	0	0	1.682
15	34	2.2	0	0	-1.682
15	34	4.8	0	0	0
15	34	4.8	0	0	0
15	34	4.8	0	0	0
15	34	4.8	0	0	0
15	34	4.8	0	0	0
15	34	4.8	0	0	0

5.4 ANALYSIS OF COVARIANCE

The analysis of covariance was obtained by fitting the following regression equation to the data:

$$\hat{y} = c_0 + c_1 I + c_1 P + c_2 S + c_3 SR + c_{11} P^2 + c_{22} S^2 + c_{33} (SR)^2 + c_{12} P \times S$$
$$+ c_{13} P \times SR + c_{23} S \times SR \qquad (5.2)$$

This equation is Equation (5.1) with the new term $c_1 I$ added. I was the initial dry matter. However, the estimates of $c_1, \ldots, c_{23}$ obtained by fitting Equation (5.2) will not be the same as those obtained by fitting Equation (5.1) because the original x-variables are correlated with the initial observations and this introduces confounding among all the x-variables. This aspect of covariance analysis is dealt with in detail in Preece (1980).

The matrix of x-variables, X, for this analysis is given in Table 5.4 and

Table 5.3 *X*-matrix used to fit Equation (5.1)

Constant x_0	P x_1	S x_2	SR x_3	P^2 x_1^2	S^2 x_2^2	$(SR)^2$ x_3^2	$P \times S$ $x_1 x_2$	$P \times SR$ $x_1 x_3$	$S \times SR$ $x_2 x_3$
1	-1	-1	-1	1	1	1	1	1	1
1	1	-1	-1	1	1	1	-1	-1	1
1	-1	1	-1	1	1	1	-1	1	-1
1	-1	-1	1	1	1	1	1	-1	-1
1	1	1	-1	1	1	1	1	-1	-1
1	1	-1	1	1	1	1	-1	1	-1
1	-1	1	1	1	1	1	-1	-1	1
1	1	1	1	1	1	1	1	1	1
1	1.682	0	0	2.828	0	0	0	0	0
1	-1.682	0	0	2.828	0	0	0	0	0
1	0	1.682	0	0	2.828	0	0	0	0
1	0	-1.682	0	0	2.828	0	0	0	0
1	0	0	1.682	0	0	2.828	0	0	0
1	0	0	-1.682	0	0	2.828	0	0	0
1	0	0	0	0	0	0	0	0	0
1	0	0	0	0	0	0	0	0	0
1	0	0	0	0	0	0	0	0	0
1	0	0	0	0	0	0	0	0	0
1	0	0	0	0	0	0	0	0	0
1	0	0	0	0	0	0	0	0	0

$(X^TX)^{-1}$ is given in Table 5.5. The analysis of variance was:

Source of variation	DF
Regression on:	
Initial	1
P	1
S	1
SR	1
P^2	1
S^2	1
$(SR)^2$	1
$P \times S$	1
$P \times SR$	1
$S \times SR$	1
Residual	9
Total	19

Equation (5.2) was fitted to the data at each harvest and the values of c_0, c_i and $c_1, \ldots, c_{23}$ obtained are given in Table 5.6. The values of $c_1, \ldots, c_{23}$ are plotted against time in Figure 5.1.

Table 5.4 **X-matrix used to fit Equation (5.2)**

Constant x_0	Initial x_i	P x_1	S x_2	SR x_3	P^2 x_1^2	S^2 x_2^2	$(SR)^2$ x_3^2	$P \times S$ $x_1 x_2$	$P \times SR$ $x_1 x_3$	$S \times SR$ $x_2 x_3$
1	7.88	1	1	1	1	1	1	1	1	1
1	7.59	1	-1	-1	1	1	1	-1	-1	1
1	8.09	-1	1	-1	1	1	1	-1	1	-1
1	8.10	-1	-1	1	1	1	1	1	-1	-1
1	7.44	1	1	-1	1	1	1	1	-1	-1
1	6.81	1	-1	1	1	1	1	-1	1	-1
1	7.47	-1	1	1	1	1	1	-1	-1	1
1	7.40	1	1	1	1	1	1	1	1	1
1	8.38	1.682	0	0	2.828	0	0	0	0	0
1	8.23	-1.682	0	0	2.828	0	0	0	0	0
1	7.85	0	1.682	0	0	2.828	0	0	0	0
1	7.10	0	-1.682	0	0	2.828	0	0	0	0
1	7.54	0	0	1.682	0	0	2.828	0	0	0
1	7.46	0	0	-1.682	0	0	2.828	0	0	0
1	7.93	0	0	0	0	0	0	0	0	0
1	7.06	0	0	0	0	0	0	0	0	0
1	7.36	0	0	0	0	0	0	0	0	0
1	7.98	0	0	0	0	0	0	0	0	0
1	7.77	0	0	0	0	0	0	0	0	0
1	7.49	0	0	0	0	0	0	0	0	0

Table 5.5 $(X^TX)^{-1}$ calculated from X given in Table 5.4

	x_0	x_i	x_1	x_2	x_3	x_1^2	x_2^2	x_3^2	$x_1 x_2$	$x_1 x_3$	$x_2 x_3$
x_0	34.58										
x_i	−4.526	0.5953									
x_1	−0.6786	0.08924	0.08660								
x_2	0.4247	−0.05585	−0.00837	0.07846							
x_3	−0.3597	0.04731	0.00709	−0.00444	0.07698						
x_1^2	0.9096	−0.12709	−0.01905	0.01192	−0.01010	0.09654					
x_2^2	−0.4188	0.04761	0.00714	−0.00447	0.00378	−0.00328	0.0732				
x_3^2	−0.3788	0.04235	0.00635	−0.00397	0.00337	−0.00216	0.01027	0.07242			
$x_1 x_2$	0.4866	−0.06399	−0.00959	0.00600	−0.00508	0.01366	−0.00512	−0.00455	0.1319		
$x_1 x_3$	−0.2376	0.03125	0.00468	−0.00293	0.00248	−0.00667	0.00250	0.00222	−0.0034	0.1266	
$x_2 x_3$	−0.0566	0.00744	0.00112	−0.00070	0.00059	−0.00159	0.00060	0.00053	−0.0008	0.0004	0.1251

5.5 REGRESSION OF THE VALUES OF $c_1, \ldots, c_{23}$ ON TIME

Ninth-degree polynomials were used initially to approximate the trends in time in the coefficients $c_1, \ldots, c_{23}$. This was the highest degree polynomial that could be fitted because there were nine degrees of freedom for error. The polynomials were fitted through the origin because all the treatment effects were expected to be zero initially. The regression equation to be fitted to each sequence of coefficients was thus:

$$\hat{c} = b_1 \xi_1 + b_2 \xi_2 + \cdots + b_9 \xi_9 \qquad (5.3)$$

where $\xi_1, \ldots, \xi_9$ are orthogonal polynomials through the origin. The values of $\xi_1, \ldots, \xi_9$ used are given in Table 5.7. Orthogonal polynomials through the origin are discussed at the end of the chapter.

The values of $b_1, \ldots, b_9$ for each plot can be calculated directly using these values and $\log_e$ (dry matter) and are given in Table 5.8. For example,

$$b_1 = [1 \times \log(5) + 2\ \log(6) + \cdots + 37\ \log(41)]/17\ 575,$$
$$b_2 = [-0.0282\ \log(5) + 0.0543\ \log(6) + \cdots + 0.3415\ \log(41)].$$

The column numbers such as (5) refer to the columns of Table 5.1.

The coefficients $b_1, b_2, \ldots, b_9$ for the regression of each of $c_1, \ldots, c_{23}$ on time were obtained indirectly by regressing the values of $b_1, \ldots, b_9$ given in Table 5.8 on the x-variables of Table 5.4. The univariate analyses of $b_1, \ldots, b_9$ are given in Table 5.9 and the error variance–covariance matrix from the multivariate analysis of $b_1, \ldots, b_9$ is given in Table 5.10. The results of the tests of significance are summarized in Table 5.11. The spaces indicate P-values greater than 0.055. Coefficients of order 6 or greater were significant at $P = 0.05$ for c_2, c_3 and c_{23}.

There were no significant time trends in the values of c_1, c_{22}, c_{33} or c_{12}. The equations for the regressions of the coefficients on time were:

S: $\quad \hat{c}_2 = 0.0019\xi_1 - 0.0131\xi_2 - 0.0546\xi_3 - 0.118\xi_4$
$\qquad\qquad + 0.0805\xi_5 - 0.189\xi_6,$ $\qquad\qquad\qquad\qquad (5.4)$

SR: $\quad \hat{c}_3 = -0.0102\xi_1 + 0.578\xi_2 - 0.115\xi_3 - 0.0162\xi_4$
$\qquad\qquad + 0.140\xi_5 - 0.329\xi_6 + 0.0971\xi_7 + 0.178\xi_8 - 0.171\xi_9,$ $\qquad (5.5)$

P^2: $\quad \hat{c}_{11} = -0.0031\xi_1,$ $\qquad\qquad\qquad\qquad\qquad\qquad\qquad (5.6)$

$P \times SR$: $\hat{c}_{13} = -0.0044\xi_1$ $\qquad\qquad\qquad\qquad\qquad\qquad\qquad (5.7)$

$S \times SR$: $\hat{c}_{23} = 0.0044\xi_1 - 0.348\xi_2 - 0.171\xi_3 - 0.0526\xi_4 - 0.172\xi_5$
$\qquad\qquad + 0.267\xi_6 - 0.315\xi_7.$ $\qquad\qquad\qquad\qquad\qquad (5.8)$

The coefficients in these regressions are the estimates given in the lower half of Table 5.9. The polynomial coefficients for each plot were also analysed omitting S^2, $(SR)^2$ and $P \times S$ from Equation (5.1) and hence from

Table 5.6   Estimates of the coefficients of Equation (5.2) for 37 harvests

	1	2	3	4	5	6	7	8	9
Constant term c_0	2.080	4.460	8.100	0.960	2.779	0.882	0.335	1.183	6.302
Initial c_i	0.7675	0.4411	−0.0361	0.8584	0.5946	0.7462	0.8348	0.9918	0.0362
P (c_1)	−0.0595	0.0804	−0.0555	−0.0932	−0.1182	0.0981	−0.1092	0.0319	−0.0680
S (c_2)	0.2012	0.0748	0.1098	−0.0427	0.0811	0.0535	−0.0834	0.0830	0.0305
SR (c_3)	0.0374	0.0554	−0.0177	0.0200	0.0054	−0.1396	−0.1311	−0.5432	−0.2781
P^2 (c_{11})	−0.1546	−0.0486	0.0825	−0.1208	−0.0720	−0.0703	0.0206	−0.0490	0.0629
S^2 (c_{22})	−0.0503	−0.0556	−0.1220	−0.0029	0.0537	0.2322	0.0584	0.1158	0.1732
SR^2 (c_{33})	0.0031	0.0214	−0.1720	−0.0565	−0.0067	0.1985	0.3021	0.0075	0.1622
P×S (c_{12})	0.0787	0.0956	0.1293	0.1807	0.2254	0.0855	−0.3024	−0.0876	−0.0473
P×SR (c_{13})	−0.1013	0.0445	0.2332	−0.0850	0.0471	−0.0439	0.0058	−0.1016	−0.3031
S×SR (c_{23})	0.1035	−0.2018	−0.3109	−0.0030	0.0114	0.1858	0.2353	0.1389	0.1494

	20	21	22	23	24	25	26	27	28
Constant term c_0	4.634	4.915	2.733	3.928	6.551	6.083	3.520	1.336	4.205
Initial c_i	0.4066	0.3273	0.5915	0.4628	0.1388	0.2052	0.4854	0.8092	0.4760
P (c_1)	−0.0644	−0.0286	0.1075	0.0660	−0.0197	−0.0127	0.0265	0.0968	−0.0949
S (c_2)	0.1409	0.0320	0.0553	0.0137	0.0027	0.0635	0.1511	−0.0578	0.0657
SR (c_3)	−0.2700	−0.2913	−0.4829	−0.1800	−0.1105	−0.3793	−0.4303	−0.3763	−0.2092
P^2 (c_{11})	−0.1052	0.0552	−0.0972	0.0097	−0.0106	−0.0861	−0.1388	−0.2394	−0.0834
S^2 (c_{22})	−0.0217	0.0874	0.1530	0.1498	−0.0181	−0.0876	0.0196	0.0865	0.0615
SR^2 (c_{33})	−0.0252	0.0008	0.0984	0.0925	0.0464	−0.0470	0.0717	−0.0143	−0.0540
P×S (c_{12})	0.0591	0.0791	0.1352	0.0480	−0.0200	0.0666	0.1122	0.1009	−0.0642
P×SR (c_{13})	−0.2311	−0.0883	−0.1402	−0.0778	0.0069	−0.0522	−0.1662	−0.2811	−0.1398
S×SR (c_{23})	0.0820	0.1635	0.1694	0.0722	0.1432	0.2003	0.1705	0.2959	0.1518

X of Table 5.2. These variables were omitted because their coefficients did not show any significant time trends. P was retained because P×SR was needed in the model. The coefficients obtained in these analyses were very similar to those given in Equations (5.4)–(5.8) and are not given. These extra analyses were necessary because the covariate introduced confounding into the design. With confounding in the design, the values obtained for the effects depend on which variables are included in the model.

The fitted regressions are shown in Figure 5.2. The regressions for c_2, c_3 and c_{23} resemble periodic curves. It may be appropriate to fit variables such as cos and sin in place of the high-order polynomials. However, the data

Harvest 10	11	12	13	14	15	16	17	18	19
7.520	6.368	−4.030	−0.969	3.059	3.709	5.894	5.130	6.272	5.579
0.0106	0.1793	1.4645	1.0834	0.5299	0.5104	0.2929	0.3766	0.2023	0.3110
0.0047	−0.0512	0.0301	0.0633	−0.0172	0.0948	0.0350	−0.0220	0.0855	0.0200
−0.0352	0.0912	−0.1689	−0.1448	0.1640	−0.1304	0.0625	0.0399	0.0868	0.0891
−0.2315	−0.3558	−0.2381	−0.4050	−0.4112	−0.2061	−0.2276	−0.1728	−0.1688	−0.1197
0.0234	−0.0743	−0.3667	−0.3252	−0.0438	−0.1812	−0.1045	−0.0506	−0.0387	−0.1164
−0.0294	−0.1077	0.1308	0.0127	0.1482	0.1501	0.0100	−0.0512	0.0230	0.0665
−0.0220	−0.0987	0.1180	0.0585	−0.0153	0.0291	0.0096	0.0673	0.0033	0.0465
0.0489	0.2246	0.2169	−0.0260	0.0111	−0.1147	0.0377	−0.0440	0.0673	0.0928
−0.0350	−0.2493	−0.1547	0.0489	−0.2006	−0.0043	−0.0604	−0.0691	−0.0950	−0.0958
0.1366	0.1501	0.1075	−0.0716	0.2258	0.0138	0.1180	0.0584	0.0739	0.2694

Harvest 29	30	31	32	33	34	35	36	37
0.474	6.071	3.852	7.759	5.846	3.385	4.641	3.059	0.508
0.9073	0.2070	0.4939	0.0306	0.2601	0.5632	0.3930	0.5942	0.8910
0.0279	−0.0212	0.0096	0.0920	0.1159	0.0480	−0.0911	0.0947	0.1299
0.1248	−0.0117	0.1167	0.0373	0.1357	0.1804	0.0512	−0.0538	0.0221
−0.3940	−0.3016	−0.1553	−0.0477	−0.2484	−0.3478	−0.2210	−0.2012	−0.2410
−0.1023	−0.0004	−0.0147	−0.0234	−0.0435	−0.1898	−0.0218	−0.0372	−0.0820
−0.0129	0.0669	−0.0395	−0.0263	−0.0250	0.0009	−0.0427	0.0450	0.0472
0.1408	0.1091	0.0355	0.0133	0.0045	0.0240	0.0241	0.0595	0.1628
0.1701	0.0868	0.0850	−0.0075	0.0576	0.0813	0.0888	0.0201	0.1040
−0.2936	−0.1184	−0.1065	0.0151	−0.1197	−0.1690	−0.0086	−0.1032	−0.0825
0.3433	0.0394	0.0277	−0.0426	0.1256	0.3350	−0.0614	0.0246	−0.0957

were collected over three years and variables such as years × cos and years × sin would probably be required. If so, there would be no advantage in using the periodic terms in the regressions because the number of variables would not be reduced.

The expected values, $\hat{c}_i$, for these regressions were obtained by substituting the values of $\xi_1,\ldots,\xi_9$ of Table 5.7 in Equations (5.4)–(5.8). The expected values are given in Table 5.12.

Algebraically,

$$\hat{c}_i = \Xi\beta$$

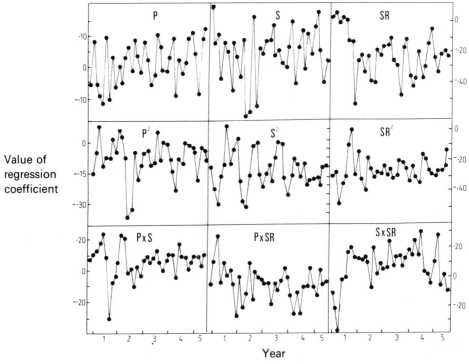

Figure 5.1   Values of c_1, c_2, c_3, c_{11}, c_{22}, c_{33}, c_{12}, c_{13}, and c_{23} of Equation (5.2) for 37 harvests.

where Ξ is a matrix made up from the columns of Table 5.7 and β is a column containing the estimated coefficients.

$$\begin{aligned}\Xi &= \xi_1 \text{ for } c_{11} \text{ and } c_{13},\\ &= \xi_1, \xi_2, \xi_3, \xi_4, \xi_5 \text{ and } \xi_6 \text{ for } c_2,\\ &= \xi_1, \ldots, \xi_7 \text{ for } c_{23},\\ &= \xi_1, \ldots, \xi_9 \text{ for } c_3.\end{aligned}$$

Then,

$$\text{var}(c_i) = \Xi \, \text{var}(\beta) \Xi^T$$

where

$$\text{var}(\beta) = x^{ii} S_b,$$

x^{ii} is the element of $(X^T X)^{-1}$ corresponding to x_i and S_b is the variance–covariance matrix of the polynomial coefficients given in Table 5.10.

Table 5.7 Values of the orthogonal polynomials used to fit Equation (5.3) to the data for each plot in Table 5.1

Harvest	ξ_1	ξ_2	ξ_3	ξ_4 ($\times 10^4$)	ξ_5	ξ_6	ξ_7	ξ_8	ξ_9
1	1	−282	632	−1109	1674	−2269	2825	−3251	4392
2	2	−543	1148	−1856	2506	−2926	2977	−2567	2960
3	3	−783	1556	−2294	2708	−2579	1843	−62	120
4	4	−1003	1863	−2473	2464	−1692	339	1138	−2031
5	5	−1202	2076	−2441	1928	−610	−980	2093	−2822
6	6	−1380	2203	−2242	1230	424	−1832	2152	−2375
7	7	−1537	2251	−1915	476	1255	−2133	1511	−1163
8	8	−1673	2228	−1498	−253	1798	−1929	493	270
9	9	−1789	2141	−1024	−892	2024	−1344	−574	1473
10	10	−1883	1998	−524	−1400	1949	−542	−1421	2165
11	11	−1957	1805	−25	−1747	1616	311	−1881	2248
12	12	−2011	1570	450	−1921	1093	1067	−1890	1779
13	13	−2043	1301	880	−1924	457	1613	−1486	930
14	14	−2055	1005	1249	−1766	−212	1877	−781	−75
15	15	−2046	689	1543	−1468	−838	1835	66	−1007
16	16	−2016	361	1751	−1058	−1353	1509	883	−1676
17	17	−1965	28	1867	−571	−1707	955	1513	−1962
18	18	−1893	−302	1887	−43	−1864	260	1837	−1830
19	19	−1801	−623	1810	486	−1809	−471	1798	−1328
20	20	−1688	−926	1641	977	−1549	−1132	1402	−572
21	21	−1554	−1205	1386	1394	−1109	−1624	722	277
22	22	−1399	−1451	1053	1701	−534	−1867	−114	1042
23	23	−1224	−1658	658	1873	116	−1816	−944	1567
24	24	−1028	−1818	215	1888	768	−1465	−1596	1743
25	25	−811	−1924	−255	1735	1345	−853	−1919	1531
26	26	−573	−1968	−728	1415	1768	−64	−1816	973
27	27	−314	−1943	−1179	940	1968	775	−1271	192
28	28	−35	−1842	−1578	339	1889	1513	−367	−631
29	29	265	−1657	−1890	−343	1506	1982	704	−1285
30	30	586	−1380	−2080	−1043	831	2035	1659	−1579
31	31	926	−1005	−2107	−1676	−74	1573	2170	−1390
32	32	1291	−525	−1929	−2137	−1080	597	1939	−723
33	33	1674	70	−1499	−2295	−1980	−740	829	251
34	34	2078	785	−766	−1992	−2469	−2059	−955	1181
35	35	2503	1628	323	−1045	−2129	−2679	−2617	1559
36	36	2949	2607	1824	760	−409	−1516	−2442	818
37	37	3415	3729	3797	3671	3396	3021	2621	−1477
Sum of squares	17 575	1	1	1	1	1	1	1	1

Table 5.8 Values of the orthogonal polynomial coefficients of order 1 to 9 calculated from the data of Table 5.1 for each plot of Lodge and Roberts (1979)

Plot	b_1	b_2	b_3	b_4	b_5	b_6	b_7	b_8	b_9
1	0.3145	−17.86	9.789	−6.729	4.569	−4.459	3.266	−2.410	2.726
2	0.3202	−17.28	8.689	−5.598	3.757	−3.462	2.676	−1.912	2.467
3	0.3123	−16.97	10.028	−6.637	5.063	−4.676	3.439	−2.295	2.752
4	0.3015	−16.54	9.501	−6.499	5.169	−4.896	3.796	−1.964	2.408
5	0.3141	−16.90	9.450	−6.239	4.950	−4.856	3.869	−2.431	3.027
6	0.2644	−14.70	8.213	−5.423	4.592	−5.602	3.456	−1.456	2.064
7	0.2944	−16.79	8.904	−6.773	4.855	−4.915	3.025	−1.850	2.397
8	0.2961	−15.85	8.370	−6.073	5.166	−5.080	3.028	−1.490	2.095
9	0.3067	−16.75	10.224	−7.023	5.402	−5.069	3.915	−2.389	2.791
10	0.3050	−16.64	9.679	−6.577	5.466	−4.883	3.674	−2.110	2.459
11	0.3123	−17.39	9.500	−6.478	4.848	−5.124	3.823	−2.153	2.561
12	0.2917	−16.51	9.415	−6.066	5.012	−4.425	3.414	−2.319	2.472
13	0.2874	−15.39	9.666	−6.447	4.956	−5.198	3.954	−2.051	2.453
14	0.3206	−17.77	9.561	−6.332	4.961	−4.182	3.484	−2.274	2.696
15	0.3105	−17.72	10.160	−6.912	5.558	−5.242	3.731	−2.282	2.689
16	0.2920	−15.08	8.429	−6.354	4.113	−5.174	3.331	−2.233	2.815
17	0.3007	−16.81	9.209	−6.202	5.052	−4.948	3.464	−2.470	2.757
18	0.3110	−16.28	8.693	−5.928	4.912	−4.883	3.954	−2.326	2.698
19	0.3045	−16.97	9.904	−6.224	4.847	−4.740	4.081	−2.574	2.787
20	0.2924	−16.63	8.466	−6.322	4.150	−5.484	4.284	−2.657	2.912

5.6 EXPECTED VALUES FOR DRY MATTER

The aims of the analysis to this stage have been to determine:

1. which variables are needed in the linear model relating dry matter and the treatment variables; and
2. how the coefficients in this linear model are related to time.

This may well be the end of the analysis for experiments such as that considered in Chapter 2 in which the treatment design was simple. Here, however, the interest is in the response of dry matter to SR, P and S and response surfaces should be examined. Expected values for dry matter are used to construct the response surfaces.

The analyses of section 5.5 indicated that dry matter could be described by the following equation:

$$\hat{y} = c_0 + c_i \bar{I} + c_1 P + c_2 S + c_3 SR + c_{11} P^2 + c_{13} P \times SR + c_{23} S \times SR. \quad (5.9)$$

The values of $\hat{y}$ can be obtained for any harvest by substituting the

Table 5.9 Analyses of variance of the orthogonal polynomial coefficients of Table 5.8 and estimates of the polynomial coefficients for each treatment variable of Equation (5.2)

(a) Analyses of variance

Source of variation	DF	Mean square ×10^6	b_1 F	P* (%)	$MS \times 10^3$	b_2 F	P (%)	$MS \times 10^3$	b_3 F	P (%)	$MS \times 10^4$	b_4 F	P (%)
Initial	1	614	46.2	0.008	2105	6.9	2.7	2267	6.1	3.5	3041	1.9	
P	1	12	<1		55	<1		1	<1		1749	1.1	
S	1	47	3.6	9.2		<1		38	<1		1786	1.1	
SR	1	1360	102.3	<0.1	4351	14.3	0.4	172	<1		34	<1	
P^2	1	97	7.3	2.4	537	1.8		75	<1		63	<1	
S^2	1	18	1.4		488	1.6		87	<1		193	<1	
$(SR)^2$	1	40	3.0		35	<1		217	<1		<1	<1	
P×S	1	40	3.0		24	<1		12	<1		652	<1	
P×SR	1	148	11.1	0.8	337	1.1		11	<1		5	<1	
S×SR	1	153	11.5	0.8	967	3.2		233	<1		221	<1	
Residual	9	13.33			304.5			370.9			1628		

*Spaces indicate P > 10.0%.

(b) Estimates and standard errors

Variable	$b_1 \times 10^3$	SE	$b_2 \times 10$	SE	$b_3 \times 10$	SE	$b_4 \times 10$	SE
P	1.03	1.07	0.69	1.62	−0.11	1.79	1.23	1.19
S	1.92	1.02	−0.13	1.54	−0.54	1.71	−1.18	1.13
SR	−10.23	1.01	5.79	1.53	−1.15	1.69	−0.16	1.12
P^2	−3.06	1.13	2.28	1.71	−0.85	1.89	−0.25	1.25
S^2	1.16	0.98	−1.89	1.49	0.80	1.65	0.38	1.09
$(SR)^2$	1.71	0.98	−0.50	1.48	1.25	1.64	<0.01	1.08
P×S	2.30	1.32	−0.57	2.00	0.41	2.21	0.93	1.47
P×SR	−4.37	1.30	2.07	1.96	0.38	2.17	0.08	1.44
S×SR	4.37	1.29	−3.48	1.95	−1.71	2.15	−0.53	1.43

Table 5.9   cont'd

(a) Analyses of variance

Source of variation	DF	b5 MS×10⁴	b5 F	b5 P (%)	b6 MS×10⁴	b6 F	b6 P (%)	b7 MS×10⁴	b7 F	b7 P (%)	b8 MS×10⁴	b8 F	b8 P (%)	b9 MS×10⁴	b9 F	b9 P (%)
Initial	1	3288	1.4		922	1.2		3283	2.9		799	1.1		82	<1	
P	1	94	<1		<1	<1		411	<1		63	<1		3	<1	
S	1	826	<1		4 566	5.9	3.8	56	<1		37	<1		304	<1	
SR	1	2558	1.1		14041	18.1	0.2	1226	1.1		4107	5.8	3.9	3814	11.9	0.7
P²	1	411	<1		<1			3371	2.9		2507	3.6	9.2	707	2.2	
S²	1	28	<1		2 585	3.3		1755	1.5		1299	1.8		1388	4.3	6.8
(SR)²	1	59	<1		3 778	4.9	5.5	916	<1		2166	3.1		899	2.8	
P×S	1	1723	<1		833	1.1		1145	<1		421	<1		314	1.0	
P×SR	1	888	<1		3 109	4.0	7.7	<1	<1		211	<1		493	1.5	
S×SR	1	2376	1.0		5 715	7.3	2.4	7915	6.9	2.7	269	<1		389	1.2	
Residual	9	2318			777.8			1146			704.8			321.7		

(b) Estimates and standard errors

Variable	b5×10³	SE	b6×10	SE	b7×10	SE	b8×10	SE	b9×10	SE
P	−0.29	1.42	0.08	0.82	0.60	1.00	0.23	0.78	0.05	0.53
S	0.81	1.35	−1.89	0.78	0.21	0.95	0.17	0.74	0.49	0.50
SR	1.40	1.34	−3.29	0.77	0.97	0.94	1.78	0.74	−1.71	0.50
P²	0.63	1.50	−0.02	0.87	−1.80	1.05	1.56	0.82	−0.83	0.56
S²	0.14	1.30	1.37	0.75	−1.13	0.92	0.97	0.72	−1.01	0.48
(SR)²	0.21	1.29	1.65	0.75	−0.81	0.91	1.25	0.71	−0.81	0.48
P×S	1.51	1.75	−1.05	1.01	1.23	1.23	−0.75	0.96	0.64	0.65
P×SR	1.06	1.71	−1.98	0.99	0.02	1.21	0.52	0.95	−0.79	0.64
S×SR	−1.72	1.70	2.67	0.99	−3.15	1.20	0.58	0.94	−0.70	0.63

Table 5.10 Error variance–covariance matrix from analysis of $b_1, \ldots, b_9$ of Table 5.8

	$b_1 \times 10^4$	b_2	b_3	b_4	b_5	b_6	b_7	b_8	b_9
b_1	0.1333								
b_2	2.332	0.3045							
b_3	−1.439	−0.1663	0.3709						
b_4	6.285	0.0812	−0.1669	0.1628					
b_5	5.664	−0.1384	0.1753	−0.0614	0.2318				
b_6	7.164	0.0540	−0.0158	0.0626	0.0088	0.0778			
b_7	−9.970	−0.0218	0.0360	−0.0200	−0.0130	−0.0507	0.1146		
b_8	6.672	0.0161	−0.0850	0.0543	−0.0030	0.0228	−0.0619	0.0705	
b_9	−4.393	0.0070	0.0459	−0.0435	−0.0181	−0.0194	0.0295	−0.0426	0.0322

Table 5.11

Regression coefficient (variable)	Polynomial coefficient								
	b_1	b_2	b_3	b_4	b_5	b_6	b_7	b_8	b_9
			P-values from Table 5.9						
$c_1(P)$	<0.001								
$c_2(S)$	0.024	0.004							
$c_3(SR)$									
$c_{11}(P^2)$					0.038			0.039	
$c_{22}(S^2)$					0.002				0.007
$c_{33}((SR)^2)$						0.055			
$c_{12}(P \times S)$									
$c_{13}(P \times SR)$	0.009								
$c_{23}(S \times SR)$					0.024		0.027		

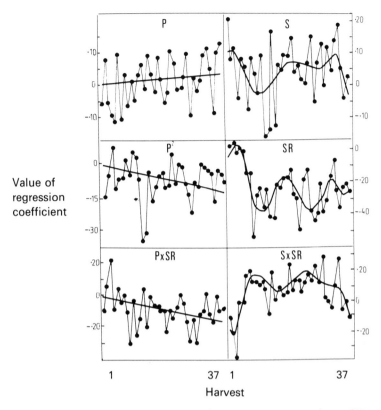

Figure 5.2   Regressions fitted to the values of $c_1, c_2, c_3, c_{11}, c_{13}$ and c_{23} of Equation (5.2) at 37 harvests. The fitted values were calculated from Equations (5.4)–(5.8).

following in Equation (5.9):

1. c_0 and c_i for that harvest;
2. $\bar{I}$, the mean initial dry matter;
3. the expected values of $c_1, c_2, c_3, c_{11}, c_{13}$ and c_{23} given in Table 5.12;
4. the values of P, S, SR, P^2, $P \times SR$ and $S \times SR$ given in Table 5.2.

If more detail is required, other values of P, S and SR can be used but they should be chosen within the range used in the experiment.

However, the effects of P, S and SR are more easily displayed as deviations from the mean dry matter for each harvest. These deviations are of the form:

$$\hat{y} - c_0 - c_i \bar{I} = c_1 P + c_2 S + c_3 SR + c_{11} P^2 + c_{13} P \times SR + c_{23} S \times SR. \quad (5.10)$$

For example, the deviations for harvests 17, 26 and 34 which were made at

Table 5.12 Expected values for c_1, c_2, c_3, c_{11}, c_{13} and c_{23} of Equation (5.2) for 37 harvests

Harvest	$c_1 \times 10$	$c_2 \times 10$	c_3	c_{11}	c_{13}	c_{23}
1	0.010	0.684	−0.040	−0.003	−0.004	−0.169
2	0.020	0.959	0.002	−0.006	−0.009	−0.197
3	0.030	0.961	0.037	−0.009	−0.013	−0.148
4	0.040	0.800	0.032	−0.012	−0.017	−0.065
5	0.050	0.559	−0.017	−0.016	−0.022	0.022
6	0.060	0.298	−0.098	−0.019	−0.026	0.096
7	0.070	0.060	−0.192	−0.022	−0.030	0.148
8	0.080	−0.129	−0.279	−0.025	−0.034	0.176
9	0.090	−0.254	−0.346	−0.028	−0.039	0.182
10	0.100	−0.311	−0.385	−0.031	−0.043	0.171
11	0.110	−0.305	−0.393	−0.034	−0.047	0.150
12	0.120	−0.243	−0.375	−0.037	−0.052	0.122
13	0.130	−0.140	−0.336	−0.040	−0.056	0.096
14	0.140	−0.008	−0.288	−0.043	−0.060	0.075
15	0.150	0.136	−0.239	−0.046	−0.064	0.062
16	0.160	0.278	−0.199	−0.050	−0.069	0.059
17	0.170	0.408	−0.173	−0.053	−0.073	0.066
18	0.180	0.514	−0.166	−0.056	−0.077	0.083
19	0.190	0.591	−0.178	−0.059	−0.082	0.105
20	0.200	0.635	−0.206	−0.062	−0.086	0.131
21	0.210	0.649	−0.245	−0.065	−0.090	0.156
22	0.220	0.635	−0.288	−0.068	−0.095	0.179
23	0.230	0.600	−0.327	−0.071	−0.099	0.196
24	0.240	0.556	−0.356	−0.074	−0.103	0.205
25	0.250	0.512	−0.369	−0.078	−0.108	0.205
26	0.260	0.481	−0.363	−0.081	−0.112	0.196
27	0.270	0.473	−0.341	−0.084	−0.116	0.180
28	0.280	0.497	−0.303	−0.087	−0.120	0.160
29	0.290	0.557	−0.260	−0.090	−0.125	0.140
30	0.300	0.651	−0.219	−0.093	−0.129	0.121
31	0.310	0.768	−0.192	−0.096	−0.133	0.109
32	0.320	0.889	−0.185	−0.099	−0.138	0.103
33	0.330	0.978	−0.203	−0.102	−0.142	0.103
34	0.340	0.983	−0.239	−0.105	−0.146	0.100
35	0.350	0.833	−0.276	−0.108	−0.151	0.082
36	0.360	0.435	−0.288	−0.112	−0.155	0.024
37	0.370	−0.332	−0.189	−0.115	−0.159	−0.108

the end of 1973, 1974 and 1975 were calculated as:

P	S	SR	P²	P × SR	S × SR
−1	−1	−1	1	1	1
1	−1	−1	1	−1	1
−1	1	−1	1	1	−1
−1	−1	1	1	−1	−1
1	1	−1	1	−1	−1
1	−1	1	1	1	−1
−1	1	1	1	−1	1
1	1	1	1	1	1
1.682	0	0	2.828	0	0
−1.682	0	0	2.828	0	0
0	1.682	0	0	0	0
0	−1.682	0	0	0	0
0	0	1.682	0	0	0
0	0	−1.682	0	0	0
0	0	0	0	0	0

$\times$

Harvest		
17	26	34
0.017	0.026	0.034
0.041	0.048	0.098
−0.173	−0.363	−0.239
−0.053	−0.081	−0.105
−0.073	−0.112	−0.146
0.066	0.196	0.100

$=$

Harvest		
17	26	34
0.055	0.292	−0.044
0.235	0.568	0.316
0.005	−0.004	−0.048
−0.277	−0.602	−0.430
0.185	0.272	0.312
−0.389	−0.774	−0.654
−0.063	−0.114	−0.034
−0.175	−0.286	−0.258
−0.121	−0.185	−0.240
−0.178	−0.272	−0.354
0.069	0.081	0.165
−0.069	−0.081	−0.165
−0.291	−0.611	−0.402
0.291	0.611	0.402
0	0	0

.

The following additional deviations were calculated in the same way. They were needed to form the response surfaces given in Figures 5.3 and 5.4.

| | | | | | | Harvest | | |
| P | S | SR | P² | P × SR | S × SR | 17 | 26 | 34 |

$$
\begin{bmatrix}
-1 & -1 & 0 & 1 & 0 & 0 \\
-1 & 0 & -1 & 1 & 1 & 0 \\
-1 & 0 & 0 & 1 & 0 & 0 \\
-1 & 0 & -1 & 1 & 1 & 0 \\
-1 & 1 & 0 & 1 & 0 & 0 \\
0 & -1 & -1 & 0 & 0 & 1 \\
0 & -1 & 0 & 0 & 0 & 0 \\
0 & -1 & 1 & 0 & 0 & -1 \\
0 & 0 & -1 & 0 & 0 & 0 \\
0 & 0 & 1 & 0 & 0 & 0 \\
0 & 1 & -1 & 0 & 0 & -1 \\
0 & 1 & 0 & 0 & 0 & 0 \\
0 & 1 & 1 & 0 & 0 & 1 \\
1 & -1 & 0 & 1 & 0 & 0 \\
1 & 0 & -1 & 1 & -1 & 0 \\
1 & 0 & 0 & 1 & 0 & 0 \\
1 & 0 & 1 & 1 & 1 & 0 \\
1 & 1 & 0 & 1 & 0 & 0
\end{bmatrix}
\begin{bmatrix}
0.017 & 0.026 & 0.034 \\
0.041 & 0.048 & 0.098 \\
-0.173 & -0.363 & -0.239 \\
-0.053 & -0.081 & -0.105 \\
-0.073 & -0.112 & -0.146 \\
0.066 & 0.196 & 0.100
\end{bmatrix}
$$

Harvest

| | 17 | 26 | 34 |

$$
=
\begin{bmatrix}
-0.111 & -0.155 & -0.237 \\
0.030 & 0.144 & -0.046 \\
-0.070 & -0.107 & -0.139 \\
0.030 & 0.144 & -0.046 \\
-0.029 & -0.059 & -0.041 \\
0.198 & 0.511 & 0.241 \\
-0.041 & -0.048 & -0.098 \\
-0.280 & -0.607 & -0.437 \\
0.173 & 0.363 & 0.239 \\
-0.173 & -0.363 & -0.239 \\
0.148 & 0.215 & 0.237 \\
0.041 & -0.048 & 0.098 \\
-0.066 & -0.119 & -0.041 \\
-0.077 & -0.103 & -0.169 \\
0.210 & 0.420 & 0.314 \\
-0.036 & -0.055 & -0.071 \\
-0.282 & -0.530 & -0.456 \\
0.005 & -0.007 & 0.027
\end{bmatrix}.
$$

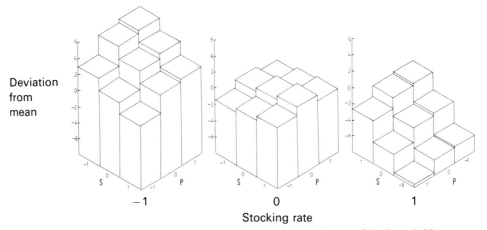

Figure 5.3 Deviations from mean dry matter at three levels of S, P and SR at harvest 26; note the change in direction of the axes for S and P at SR 1.

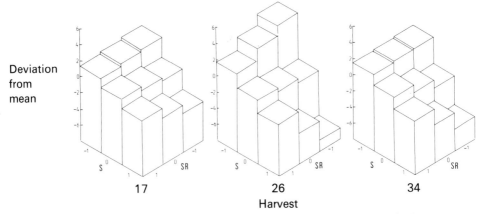

Figure 5.4 Deviations from mean dry matter at three levels of S and SR for harvests 17, 26 and 34. The deviations were averaged over P-levels of −1, 0 and 1.

5.7 POLYNOMIALS THROUGH THE ORIGIN

Treatment contrasts can be assumed to be zero initially in a well-designed experiment. Trends fitted to the values of the contrasts should allow for this. With polynomials this is conveniently done by using polynomials through the origin. The linear polynomial multipliers for polynomials through the origin are always the selected values of the variable, i.e. $\xi_{1,i}=t_i$. The quadratic multipliers can always be calculated as:

$$\xi_{2,i}=\lambda(t_i^2-kt_i)$$

where
$$k=\sum t_i^3/\sum t_i^2$$

and λ is a convenient multiplier. For equally spaced intervals,

$$k = 3N(n+1)/2(2n+1)$$

where n is the number of intervals. The linear and quadratic polynomials are often adequate to approximate the main features of sequences of treatment contrasts.

A recursive method of producing orthogonal polynomials through the origin is given in Appendix 2 of Green and Margerison (1977). Computer programs that generate the polynomials are generally available. Values of polynomials through the origin have not been extensively tabled. Values of the polynomials up to order 10 for equally spaced intervals up to $n=24$ are given in Appendix 1. The values of the polynomials quickly become large and the sums of squares much larger. It is convenient to use the values divided by their sum of squares, i.e. standardized values, although the linear polynomial is naturally convenient.

If there is doubt that polynomials through the origin are appropriate, a test can be carried out. The intercept, a, is calculated for each plot using ordinary polynomials or the values of $t, t^2, \ldots, t^n$. The values of a are analysed using the model appropriate to the experiment to test the hypothesis that the intercept is zero for each treatment contrast. However, no test should be needed in a well-designed experiment. Design is considered in Chapter 6.

5.8 DISCUSSION

The analysis of section 5.5 is equivalent to fitting regressions to the values of $c_1, \ldots, c_{23}$ separately. This is not strictly appropriate for non-orthogonal designs, including orthogonal designs in which covariates have been introduced. The appropriate analysis would fit a linear model to all the data together. Here, the time-variables $\xi_1, \ldots, \xi_7$ and their interactions with P, S, SR, P^2, S^2, $(SR)^2$, $P \times S$, $P \times SR$ and $S \times SR$ would be added to Equation (5.2) and the model fitted, allowing for a mixture of independent and correlated errors. This analysis requires methods that are beyond the scope of this book and, in fact, the method has only been developed recently (Verbyla and Cullis, submitted for publication).

Preece (1980) investigated the problem of non-orthogonality induced in factorial experiments by covariates in the usual case of independent errors. He concluded on p. 121 'that induced non-orthogonality may not be very important in practice, at least for well-designed experiments whose x-variate is free of large mean squares for d.f. for treatments.' It seems likely that this conclusion will also apply to experiments such as that considered here, in which errors between some of the observations are correlated. This problem in the analysis can often be avoided by using the information on the covariate in the planning state; this aspect is considered in section 6.2.4.

6 Pre-treatment observations in the design of experiments

6.1 INTRODUCTION

Response variables that can be measured repeatedly during an experiment can often also be measured before the treatments are applied, e.g. live weights, heights of trees in the nursery, hormone levels in blood and composition and yield of pasture. These observations can be very important at the planning stage by ensuring that a suitable experimental design is chosen.

Pre-treatment observations (x) are used mainly in two ways:

1. to increase the precision of the experiment;
2. to ensure that comparisons are not biased through the way the individuals are allocated to treatments.

In most biological experiments the choice will be between:

1. using the values of x to form blocks; or
2. using x as a covariate for later observations.

Covariance analysis was considered in Chapter 5.

6.2. BLOCKING

The values given in Table 6.1 are the initial average heights (cm) of 27 groups of four trees to be used to compare three treatments A, C and D for the control of the insect, leaf miner, in young oranges. Height was considered to be an appropriate measurement because leaf miner affects the shoots. The heights have been arranged in descending order in columns of three and some totals needed later have been added. The variation among these 27 values gave $s^2 = 23.56$ on 26 degrees of freedom.

6.2.1. Simple blocks

Each column could be regarded as a block and the treatments applied at random in each. One such allocation is given in Table 6.2, from which Table 6.3 was formed.

Table 6.1

| | Column | | | | | | | | | Row |
	1	2	3	4	5	6	7	8	9	totals
	80	75	73	71	69	68	67	66	64	633
	79	74	72	70	69	68	67	66	62	627
	76	73	72	69	68	67	66	65	59	615
Column totals	235	222	217	210	206	203	200	197	185	1875 (grand total)

Table 6.2

| | | | | Column | | | | |
1	2	3	4	5	6	7	8	9
80D	75D	73A	71C	69A	68C	67A	66C	64D
79A	74C	72C	70D	69D	68D	67D	66D	62A
76C	73A	72D	69A	68C	67A	66C	65A	59C

Table 6.3

| Treatment | | | | Column | | | | | | Treatment |
	1	2	3	4	5	6	7	8	9	totals
A	79	73	73	69	69	67	67	65	62	624
C	76	74	72	71	68	68	66	66	59	620
D	80	75	72	70	69	68	67	66	64	631
	235	222	217	210	206	203	200	197	185	1875

Then,

$$\text{total sum of squares} = 79^2 + 73^2 + \cdots 66^2 + 64^2 - 1875^2/27$$
$$= 612.67,$$

$$\text{blocks sum of squares} = (235^2 + 222^2 + \cdots + 197^2 + 185^2)/3$$
$$- 1875^2/27$$
$$= 584.03,$$

$$\text{and treatments sum of squares} = (624^2 + 620^2 + 631^2)/9 - 1875^2/27$$
$$= 6.89.$$

The residual variance has been reduced from 23.564 on 26 degrees of freedom to 1.359 on 16 degrees of freedom by allowing for blocks.

This allocation produced the analysis of variance of initial heights given in Table 6.4.

Table 6.4

Source	DF	SS	MS
Blocks	8	584.03	
Treatments	2	6.89	3.445
Residual	16	21.75	1.359
Total	26	612.67	23.564

6.2.2 Allowing for order within blocks

In this example the treatments can be allocated to initial height, taking into account the ordering of x within each block as well as the differences between blocks. This can be done whenever the number of blocks is a multiple of the number of treatments. The 3×9 arrangement of Table 6.1 can be considered as three 3×3 row–column squares. Allocating treatments to each square as in a 3×3 latin square ensures that each treatment occurs three times in each row as well as once in each column. An example of such an allocation is given in Table 6.5.

In this arrangement the treatments × columns table was as given in Table 6.6. The appropriate analysis of variance was then as shown in Table 6.7.

Table 6.5

				Column				
1	2	3	4	5	6	7	8	9
C	A	D	A	C	D	D	C	A
A	D	C	C	D	A	C	A	D
D	C	A	D	A	C	A	D	C

Table 6.6

Treatment	Column									Treatment totals
	1	2	3	4	5	6	7	8	9	
A	79	75	72	71	68	68	66	66	64	629
C	80	73	72	70	69	67	67	66	59	623
D	76	74	73	69	69	68	67	65	62	623

Table 6.7

Source	DF	SS	MS
Blocks (columns)	8	584.03	
Rows	2	18.68	
Treatments	2	2.67	1.335
Residual	14	7.30	0.521
Total	26	612.67	

The sums of squares for total and blocks were the same as before and

$$\text{rows sum of squares} = (633^2 + 627^2 + 615^2)/9 - 1875^2/27,$$
$$= 18.67$$

and treatment sum of squares $= (629^2 + 623^2 + 623^2)/9 - 1875^2/27$
$$= 2.67$$

The residual variance has been further reduced to 0.521 on 14 degrees of freedom by allowing for order within blocks.

6.2.3 Blocking on an additional criterion

Blocking can be based on other criteria, as well as on the values of the concomitant variable. For example, location in the field or in a livestock house or glasshouse often influences growth and yield and very worthwhile gains in precision have been obtained by blocking on location. The values in Table 6.8 are the total heights of two-tree plots initially available for a 3^4 factorial experiment on the effects of N, P, K and lime L on young oranges.

The rows were contour rows and formed a natural blocking system. The intention was to use a design in which the 81 treatments were laid out in blocks of nine as in Plan 6.8 of Cochran and Cox (1957). However, inspection supported by measurements showed considerable variation in initial size. The initial heights were arranged in order of magnitude within each row (block) as in Table 6.9.

The variation in initial height was then divided as in Table 6.10, where

$$\text{total sum of squares} = 152^2 + 147^2 + \cdots + 105^2 + 104^2$$
$$- 11\,112^2/81,$$
$$\text{between rows sum of squares} = (1260^2 + \cdots + 1094^2)/9 - 11\,112^2/81$$

and between strata sum of squares $= (1418^2 + \cdots + 1033^2)/9 - 11\,112^2/81.$

Table 6.8 Initial heights (cm) of two-tree plots laid out in nine contour rows

Row	Position in row								
	1	2	3	4	5	6	7	8	9
1	143	162	143	133	130	132	145	135	147
2	136	154	132	132	130	130	112	126	120
3	136	146	142	159	164	145	135	120	128
4	139	127	155	149	110	139	137	153	134
5	130	162	167	162	130	159	142	137	110
6	128	125	150	108	154	133	157	163	136
7	142	152	118	147	139	130	124	142	143
8	141	143	151	121	125	142	160	143	152
9	112	151	121	125	114	105	138	124	104

Table 6.9   Heights of Table 6.8 arranged in nine strata formed on their magnitude within each row

| Row | Height strata | | | | | | | | | Totals |
	1	2	3	4	5	6	7	8	9	
1	162	147	145	143	143	135	133	132	130	1 260
2	154	136	132	132	130	130	126	120	112	1 172
3	164	159	146	145	142	136	135	128	120	1 275
4	155	153	149	139	139	137	134	127	110	1 243
5	167	162	162	159	142	137	130	130	110	1 299
6	163	157	154	150	136	133	128	125	108	1 254
7	152	147	143	142	142	139	130	124	118	1 237
8	160	152	151	143	143	142	141	125	121	1 278
9	151	138	125	124	121	114	112	105	104	1 094
Totals	1418	1351	1307	1277	1238	1203	1169	1116	1033	11 112

Table 6.10

	DF	SS	MS
Between rows	8	3 605.3	
Between strata	8	12 694.0	
Residual	64	1 646.9	25.73
Total	80	17 946.2	

This analysis indicated that blocking on initial height should be important because rows were also to be used as blocks. A 9×9 quasi-latin square (Cochran and Cox, 1957, p. 331) was considered suitable.

Square II of Cochran and Cox was selected on the toss of a coin. The rows and the columns of this square were permuted to give the layout in Table 6.11. Each set of four integers indicates the levels of N, P, K and L

Table 6.11   Allocation of treatments in a 3^4 factorial to rows and height strata

| Row | Height strata | | | | | | | | |
	1	2	3	4	5	6	7	8	9
1	2012	1102	1211	0110	0001	2200	2121	1020	0222
2	0021	2111	2220	1122	1010	0212	0100	2002	1201
3	1222	0012	0121	2020	2211	1110	1001	0200	2102
4	1111	0201	0010	2212	2100	1002	1220	0122	2021
5	0102	2222	2001	1200	1121	0020	0211	2110	1012
6	1000	0120	0202	2101	2022	1221	1112	0011	2210
7	2201	1021	1100	0002	0220	2122	2010	1212	0111
8	2120	1210	1022	0221	0112	2011	2202	1101	0000
9	0210	2000	2112	1011	1202	0101	0022	2221	1120

making up the treatment to be applied; the levels of each variable were coded 0, 1, 2.

6.2.4 Confounding

The use of pre-treatment observations to form blocks can produce a simpler analysis than using them as covariates. This is particularly important with complex experiments. For example, all effects were confounded in the experiment of Lodge and Roberts (1979) considered in Chapter 5 when the covariate was introduced. However, most of the variation caused by differences in initial pasture available could have been removed without introducing confounding by blocking on the values of initial dry matter, using Plan 8A.4 of Cochran and Cox (1957). To illustrate this design, the six plots with the highest pasture were assigned to Block I, the next six to Block II and the remainder to Block III. The treatments were randomized within these blocks.

Blocking divided the variation in initial pasture as in Table 6.12. The block totals were 43.17, 45.32 and 64.44 with total 152.93 so that the blocks sum of squares was

$$43.17^2/6 + 45.32^2/6 + 64.44^2/8 - 152.92^2/20.$$

The residual mean square was reduced from 0.1693 to 0.0357 by blocking.

The resulting design is given in Table 6.13. The appropriate analysis of variance after the treatments were allocated to the plots was:

Source	DF
Blocks	2
P	1
S	1
SR	1
P^2	1
S^2	1
$(SR)^2$	1
$P \times S$	1
$P \times SR$	1
$S \times SR$	1
Residual	8
Total	19

Table 6.12

	DF	SS	MS
Between blocks	2	2.610172	
Residual	17	0.606283	0.0357
Total	19	3.216455	0.1693

Table 6.13 Design for the experiment of Lodge and Roberts laid out in three incomplete blocks

Initial	Block	P	S	SR
6.81	1	0	0	0
7.06	1	−1	1	−1
7.10	1	1	−1	−1
7.36	1	−1	−1	1
7.40	1	1	1	1
7.44	1	0	0	0
7.46	2	1	−1	1
7.47	2	0	0	0
7.49	2	−1	1	1
7.54	2	1	1	−1
7.59	2	0	0	0
7.77	2	−1	−1	−1
7.85	3	0	−1.682	0
7.88	3	−1.682	0	0
7.93	3	0	1.682	0
7.98	3	0	0	0
8.09	3	0	0	1.682
8.10	3	1.682	0	0
8.23	3	0	0	0
8.38	3	0	0	−1.682

The analysis was obtained by fitting the following equation to the data:

$$\hat{y} = c_0 + c_4 B1 + c_5 B2 + c_1 P + c_2 S + c_3 SR + c_{11} P^2 + c_{22} S^2 + c_{33} (SR)^2$$
$$+ c_{12} P \times S + c_{13} P \times SR + c_{23} S \times SR \tag{6.1}$$

where B1 and B2 were variables used to estimated the effects of the blocks. The X-matrix needed to fit this equation is given in Table 6.14. Equation (6.1) is much simpler to fit than Equation (5.2) because the variables used to fit Equation (6.1) are nearly orthogonal.

Similarly the NPKL experiment on oranges could have been blocked on rows only and initial height used as a covariate. However, all treatment effects would have been confounded with the covariate and with each other to some degree. Whereas in the design used, in which initial height was used to form blocks instead of as a covariate, only second-order interaction effects were confounded with initial height. The analysis, which is straight-forward, is considered in section 6.3.

6.3. ALLOCATION OF TREATMENTS

Randomization is an essential feature of experimental design since it ensures that estimates are unbiased. However, all treatments should have about the

Table 6.14 X-matrix used to fit Equation (6.1)

B1	B2	P	S	SR	P^2	S^2	$(SR)^2$	$P \times S$	$P \times SR$	$S \times SR$
1	2	0	0	0	0	0	0	0	0	0
1	2	-1	1	-1	1	1	1	-1	1	-1
1	2	1	-1	-1	1	1	1	-1	-1	1
1	2	-1	-1	1	1	1	1	1	-1	-1
1	2	1	1	1	1	1	1	1	1	1
1	2	0	0	0	0	0	0	0	0	0
-1	2	1	-1	1	1	1	1	-1	1	-1
-1	2	0	0	0	0	0	0	0	0	0
-1	2	-1	1	1	1	1	1	-1	-1	1
-1	2	1	1	-1	1	1	1	1	-1	-1
-1	2	0	0	0	0	0	0	0	0	0
-1	2	-1	-1	-1	1	1	1	1	1	1
0	-3	0	1.682	0	0	2.828	0	0	0	0
0	-3	-1.682	0	0	2.828	0	0	0	0	0
0	-3	0	1.682	0	0	2.828	0	0	0	0
0	-3	0	0	0	0	0	0	0	0	0
0	-3	0	0	1.682	0	0	2.282	0	0	0
0	-3	1.682	0	0	2.828	0	0	0	0	0
0	-3	0	0	0	0	0	0	0	0	0
0	-3	0	0	-1.682	0	0	2.282	0	0	0

same initial mean value for the response variable so that all treatment effects are close to zero initially. This cannot be assumed in a single randomization.

For example, the analysis of the initial heights of Table 6.9 under the layout of Table 6.11 is given in Table 6.15. Two effects, K^2 and N^2L, were apparently, significant at $P=0.05$ and a third, PL^2, was nearly so. This randomization was thus not completely satisfactory. A new randomization and a new analysis would be carried out. When such a large number of parameters are to be estimated, a number of randomizations may need to be examined before a satisfactory one is obtained. In this example, only $19\% = 0.95^{32}$ of randomizations would be expected to have the estimates of all 32 parameters non-significant at $P=0.05$. However, there are a lot of suitable randomizations available. The total number of possible randomizations of this design is 9! (factorial 9) × 9! or nearly 1.32×10^{11}.

The estimated residual variance of 21.52 in the above example was close to the 25.73 found for the stratified data. This need not be so.

The initial pasture available in the experiment of Lodge and Roberts (Chapter 5) will be used to illustrate the outcome of randomization of treatments to plots. The 20 plots were allocated to three blocks on initial pasture available as in Plan 8A.4 of Cochran and Cox (1957). The residual variance was then 0.0357. Equation (6.1) was fitted to the values of initial

Table 6.15 Analysis of variance of the heights of Table 6.9 under the layout of Table 6.11

Source of variation	DF	MS	F	P (%)
Rows	8	450.7	20.9	0
Heights	8	1586.8	73.7	0
N	1	68.9	3.2	8
N^2	1	20.1	0.9	34
P	1	4.2	0.2	66
P^2	1	20.1	0.9	34
K	1	26.7	1.2	27
K^2	1	128.0	5.9	2
L	1	2.7	0.1	73
L^2	1	43.6	2.0	16
NP	1	40.1	1.9	18
NK	1	34.0	1.6	22
NL	1	1.4	0.1	80
PK	1	20.2	0.9	34
PL	1	69.4	3.2	8.2
KL	1	4.7	0.2	64
N^2P	1	8.3	0.4	54
N^2K	1	7.8	0.4	55
N^2L	1	102.1	4.7	3.7
P^2N	1	10.7	0.4	49
P^2K	1	39.1	1.8	19
P^2L	1	3.0	0.1	71
K^2N	1	46.7	2.2	15
K^2P	1	2.1	0.1	76
K^2L	1	18.8	0.9	36
L^2N	1	57.8	2.7	11
L^2P	1	85.3	4.0	5.5
L^2K	1	34.5	1.6	21
N^2P^2	1	25.0	1.2	29
N^2K^2	1	17.4	0.8	38
N^2L^2	1	0.2	0	91
P^2K^2	1	0.7	0	86
P^2L^2	1	0	0	100
K^2L^2	1	14.7	0.7	41
Error	32	21.5		

pasture for the designs given by 1000 randomizations of the treatments to plots.

As nine treatment parameters are to be estimated, $37\% = 100(1 - 0.95^9)$ of randomizations would be expected to produce at least one estimate that was apparently significant at $P = 0.05$.

Since the total variation is constant, such randomizations would be expected to give estimates of variance lower than 0.0357. Also, a similar proportion of randomizations would give estimates of residual variance considerably larger than 0.0357. This is shown in Table 6.16 which summarizes the results of analysing 1000 randomizations.

There are no rules for choosing an acceptable variance. However, it would be unnecessarily conservative to choose a randomization that had a residual variance much larger than 0.0357. This would make real effects more difficult to detect. The two classes, 0.030–0.0349 and 0.035–0.0399 in Table 6.16 are conveniently close to 0.0357; 241 or 85.8% of randomizations in these classes had no significant effects initially. One of these 24.1% of all randomizations would certainly be satisfactory. The total number of possible randomizations was again large, being $6! \times 6! \times 8!$ or nearly 2.1×10^{10}.

6.4 MORE THAN ONE CONCOMITANT

Allocation of treatments to units can be more difficult in experiments in which a number of variables are to be considered or in which the response is complex. For example, Robinson (personal communication) studied the composition of grazed pasture. The pasture which had been sown was made up of white clover, subterranean clover and grasses. The pasture was grazed continuously in five systems except that the sheep were removed during the following periods for systems (S) 1, 2, 3 and 4: S1 – winter; S2 – spring; S3 – summer; and S4 – autumn. The data were the number of times out of

Table 6.16 Estimated residual variances in 1000 randomizations of the initial pasture of Lodge and Roberts using Equation (6.1) and number of randomization with one or more effects significant at $P = 0.05$

Estimated residual variance	Frequency	Number with one or more significant effects
<0.015	28	28
0.015–0.0199	74	73
0.020–0.0249	108	90
0.025–0.0299	158	78
0.030–0.0349	147	30
0.035–0.0399	134	10
0.040–0.0449	129	0
0.045–0.0499	84	0
0.050–0.0549	63	0
0.055–0.0599	51	0
>0.0599	24	0
Total	1000	309

420 that a point on a star-like wheel touched white clover, subterranean clover, grass or bare ground as it was rolled across a plot.

The area was split into 20 plots to which the five systems were to be randomized to provide five replicates of each system. The initial counts are given in Table 6.17, together with five randomizations of the treatments to the plots. The variation between the five systems was divided into the following comparisons:

1. M1: $S1 + S2 + S3 + S4 - 4S5$;
2. M2: $S1 - S2 + S3 - S4$;
3. M3: $S1 - S3$; and
4. M4: $S2 - S4$.

M1 compared paddocks that had been rested with those that had been grazed continuously; M2 compared resting in winter or summer with resting in spring or autumn; M3 compared resting in winter with resting in summer; M4 compared resting in spring with resting in autumn. Other comparisons could have been chosen but these were selected as the most appropriate because spring and autumn were the periods of active growth in the region.

Table 6.17 Counts of four components of pasture on 20 plots and treatments allotted to the plots in five randomizations

Plot	Counts				Treatment allotted in randomizations R1...R5				
	White clover	Subterranean clover	Grasses	Bare ground	R1	R2	R3	R4	R5
1	152	142	34	92	2	3	1	4	5
2	115	90	15	200	3	5	2	3	1
3	207	30	15	168	4	2	5	5	3
4	159	49	38	174	5	4	4	2	2
5	156	130	24	110	1	1	3	1	4
6	173	60	54	133	3	2	4	2	4
7	186	54	42	138	5	4	5	4	5
8	101	64	86	169	4	5	2	5	1
9	175	68	87	90	2	3	3	1	2
10	223	47	66	84	1	1	1	3	3
11	163	93	21	143	4	5	2	2	4
12	209	60	11	140	5	4	4	1	5
13	172	76	35	137	1	3	1	3	3
14	248	49	66	57	2	2	5	5	1
15	133	30	24	233	3	1	3	4	2
16	217	56	31	116	5	2	4	2	4
17	192	90	55	83	1	4	5	5	2
18	203	60	44	113	3	5	2	3	1
19	220	59	26	115	4	3	1	4	5
20	255	80	59	26	2	1	3	1	3

Table 6.18 Analyses of variance of five randomizations of the data of Table 6.17

Source	DF	White clover			Subterranean clover			Grasses			Bare ground		
		MS	F	P (%)	MS	F	P (%)	MS	F	P (%)	MS	F	P (%)
Between plots	19	1688			830			512			2374		
					Randomization 1								
M1	1	480	0.3	61	1066	1.3	27	622	1.3	28	1271	1.1	32
M2	1	1482	0.8	37	1	0.0	99	371	0.8	40	3393	2.8	11
M3	1	1770	1.0	33	1326	1.6	22	231	0.5	50	8778	7.3*	1.6
M4	1	2415	1.4	26	1081	1.3	27	1201	2.5	14	13612	11.3*	0.43
Error	15	1728			819			487			1204		
					Randomization 2								
M1	1	7013	4.6*	4.9	274	0.3	58	1	0.0	99	4560	1.7	21
M2	1	689	0.5	51	2116	2.5	13	116	0.2	68	930	0.4	56
M3	1	288	0.2	68	421	0.5	49	10	0.0	90	45	0.0	90
M4	1	1225	0.8	38	421	0.5	49	50	0.0	78	465	0.2	68
Error	15	1524			836			636			2605		
					Randomization 3								
M1	1	3200	2.1	17	925	1.0	32	41	0.1	81	1059	0.4	53
M2	1	1332	0.9	37	625	0.7	42	189	0.3	59	5663	2.2	15
M3	1	288	0.2	67	32	0.0	85	136	0.2	64	120	0.0	83
M4	1	3872	2.5	14	841	0.9	35	128	0.2	65	480	0.2	67
Error	15	1559			889			615			2519		
					Randomization 4								
M1	1	82	0.0	84	616	0.6	43	959	1.7	21	231	0.1	77
M2	1	689	0.3	57	289	0.3	59	315	0.6	46	3721	1.5	24
M3	1	840	0.4	53	528	0.6	47	55	0.1	76	3528	1.4	25
M4	1	55	0.0	87	91	0.1	76	40	0.1	79	18	0.0	93
Error	15	2027			949			547			2508		
					Randomization 5								
M1	1	387	0.2	64	442	0.5	49	898	1.7	21	115	0.0	84
M2	1	1521	0.9	36	400	0.4	52	169	0.3	58	1024	0.4	55
M3	1	4512	2.7	12	112	0.1	73	162	0.3	59	1922	0.7	42
M4	1	312	0.2	67	1300	1.4	25	684	1.3	27	760	0.3	61
Error	15	1689			901			521			2753		

*$P < 5\%$.

The analysis of variance was then:

Source	DF
M1	1
M2	1
M3	1
M4	1
Error	15

The results of analysing the outcomes of the five randomizations are given in Table 6.18 for each species and bare ground. The variance between plots is also given in Table 6.18 for each variable. Randomization 1 was unsatisfactory because M3 and M4 were significant for bare ground. Randomization 2 was also unsatisfactory because M1 was significant for white clover. Randomization 3 was satisfactory and would have been selected since the randomizations were carried out in order.

6.5 GENERAL COMMENTS

Blocking has the very important effect of allowing the treatments to be compared with the same precision as would be expected from using much more uniform material than was actually used. Covariance has the same effect, although this is not obvious until the analyses of later observations are carried out.

The number of degrees of freedom assigned to blocks can often be reduced by combining blocking variables. Field blocks (rows) and height blocks had to be different in the NPKL experiment of section 6.2.3 because the trees were planted before blocking on size was considered. If the heights had been measured in the nursery, similar sized trees could have been planted in each row. This would have halved the number of blocks. Covariance may be preferred to blocking because it usually requires fewer degrees of freedom. However, blocking remains effective if the regression of y on x is non-linear, but covariance does not.

The use of pre-treatment observations in selecting an experimental design has been considered by Cox (1957, 1982). Fortunately, the general conclusion was that covariance was only appreciably better than the simpler method of blocking when the correlation between the initial and final observations was at least 0.8. This is high.

The increase in precision from using pre-treatment observations for blocking or as covariates can be so large that the measurements should always be taken even if they are expensive.

7 Weighted regression, goodness-of-fit and related topics

7.1 INTRODUCTION

The regression equations used in the previous chapters were selected from tests of significance of individual coefficients. The individual coefficients are correlated because the original observations were correlated. This testing procedure may then lead to selection of some coefficients that are not really needed. Such coefficients only appear important through their correlation with others. In applications of regression analysis in which the deviations from regression are used as errors, overfitting usually underestimates the error. In the earlier chapters error was estimated from the experimental design and overfitting is not likely to be so important.

The testing procedure does not provide tests of lack-of-fit of the assumed regression unless all the likely coefficients can be nominated, e.g. a polynomial of degree p. However, it does not seem possible to obtain appropriate tests of lack-of-fit without making assumptions about the distribution of the observations.

Generalized least squares and maximum likelihood produce an appropriate test but they can only be used when the distribution of the variables is known. Generalized least squares (GLS) will be considered here because of its obvious relation to ordinary least squares used in the rest of the book. Generalized least square and maximum likelihood are equivalent when the distribution is known. Direct fitting using ordinary least squares is a useful starting point.

7.2. DIRECT FITTING

The indirect method was used in the previous chapters to fit regressions to the values of the treatment comparisons. This ensured that only well-known methods of analysis were needed. However, the regressions could have been fitted to the treatment directly. Direct fitting will be considered in this chapter to introduce advanced methods that may be needed in some analyses. There are at least two stages in direct fitting also. Consider the introductory example of Chapter 1 again to fix ideas.

Stage 1

Calculate the means, $\bar{y}_i$, $i=1,\ldots,52$, and the sample variance–covariance matrix, S_y, of the observations on the ten sheep for the 52 sampling occasions. Symbolically

$$
\underset{52\times 52}{S_y} =
\begin{bmatrix}
s_1^2\ s_1 s_2 r_{12} & & & s_1 s_{52} r_{152} \\
s_1 s_2 r_{12}\ s_2^2 & & & \\
& \cdot & & \\
& & \cdot & \\
& & & \cdot \\
s_1 s_{52} r_{152} & & & s_{52}^2
\end{bmatrix}
$$

Then

$$
S_{\bar{y}} = S_y/10.
$$

Stage 2

Fit the following regression equation to the means by ordinary least squares (OLS)

$$
\hat{y}_i = b_0 + b_{11}\cos t + b_{12}\sin t \tag{7.1}
$$

to estimate b_0, b_{11} and b_{12}. These estimates are given by

$$
\hat{\boldsymbol{\beta}}_{\text{OLS}} = (\boldsymbol{P}^{\text{T}}\boldsymbol{P})^{-1}\boldsymbol{P}^{\text{T}}\boldsymbol{y}
$$
$$
= \boldsymbol{D}\boldsymbol{y}.
$$

$\boldsymbol{P}$ consists of a column of 1s and columns t_{11} and t_{12} of Table 1.7. Then

$$
\text{var}(\boldsymbol{\beta}_{\text{OLS}}) = \boldsymbol{D}\boldsymbol{S}_{\bar{y}}\boldsymbol{D}^{\text{T}}. \tag{7.2}
$$

The estimates obtained this way are the same as those calculated indirectly, as are their variances when S is non-singular.

Importantly, the elements of $\boldsymbol{\beta}_{\text{OLS}}$ are always unbiased estimates of b_0, b_{11} and b_{12} of Equation (7.1). However, they only have minimum variance when the observations taken on each occasion are uncorrelated and have the same variance. When this is not so, the generalized least squares estimates, to be defined now, have minimum variance. However, these can only be obtained when the variances and covariances of the observations are known.

7.3 GENERALIZED LEAST SQUARES

The generalized least squares estimates are given by

$$\beta_{GLS} = (P^T \Sigma^{-1} P)^{-1} P^T \Sigma^{-1} Y \qquad (7.3)$$

where Σ is the known variance–covariance matrix of the observations. These estimates are also unbiased and their variance is $P^T \Sigma^{-1} P$. β_{GLS} are more efficient estimates than β_{OLS} in the sense that they have lower variance in general.

The analysis of variance for the 52 mean thyroxine values based on generalized least squares estimates of the three parameters of Equation (7.1) would take the form of Table 7.1. This is slightly different from the usual analysis in that the mean is included in the analysis of variance table. This is necessary because the mean is non-orthogonal to the other parameters estimated by generalized least squares. The regression and deviations sums of squares are distributed as χ^2 because Σ is assumed to be known. The deviations provide the test of lack-of-fit.

Table 7.1

Source	DF	SS
Mean	1	$\beta_0 P^T \Sigma^{-1} y$
Regression	2	$\beta P^T \Sigma^{-1} y - \beta_0 P^T \Sigma^{-1} y$
Deviations	49	difference
Total	52	$y^T \Sigma^{-1} y$

7.4 WEIGHTED LEAST SQUARES USING ESTIMATED WEIGHTS

Unfortunately, Σ is always unknown in biological experiments. However, it seems sensible to try to take advantage of a possible decrease in variance by using an estimate, V, of Σ in Equation (7.3). S is not usually a useful estimate of Σ. S is singular when the number of coefficients in the regression exceeds the number of degrees of freedom for error, hence S^{-1} does not exist. Even when S is non-singular it is not a good estimate because a large number of parameters are estimated in it. A lot of research is currently in the literature on modelling Σ and calculating V; see, for example, Verbyla and Cullis (1990). Estimates of variance components and aspects of time-series analysis are involved and are beyond the scope of this book. Standard methods of weighted regression are used once V has been calculated.

Some aspects of weighted least squares are important for a general understanding of the analysis of the type of data considered here. A simple example will be considered now to illustrate these, although the methods are not used in this book. The estimate of Σ will be obtained in an *ad hoc* manner to illustrate the method.

The general form of Σ is:

$$
\begin{bmatrix}
\sigma_1^2 & \rho_{12}\sigma_1\sigma_2 & \cdot & \cdot & & \rho_{1p}\sigma_1\sigma_p \\
\rho_{12}\sigma_1\sigma_2 & \sigma_2^2 & & & & \\
& & & & \cdot & \\
& & & \cdot & & \\
& & \cdot & & & \\
\rho_{1p}\sigma_1\sigma_p & & & & & \sigma_p^2
\end{bmatrix}
$$

If the variances and covariances can be assumed to be the same over the period, Σ can be simplified to

$$
= \sigma^2
\begin{bmatrix}
1 & \rho_1 & \cdot & \cdot & \cdot & \rho_{p-1} \\
\rho_1 & & \cdot & & & \\
& & & \cdot & & \\
& & & & \cdot & \\
\rho_{p-1} & & & & & \rho_{11}
\end{bmatrix}
$$

where ρ_1 is the correlation coefficient of lag 1, ρ_2 that of lag 2 and so on. Then S can be constructed to have the following similar form:

$$
S = s_m^2
\begin{bmatrix}
1 r_1 & & \cdot & \cdot & \cdot & r_{p-1} \\
r_1 & \cdot & & & & \\
& & \cdot & & & \\
& & & \cdot & & \\
r_{p-1} & & & & & 1
\end{bmatrix}
$$

$$
= s_m^2 R'
$$

where, s_m^2 is the mean of the p estimates of variance, r_1 is the average of the $p-1$ correlation coefficients of lag 1, r_2 is the average of the $p-2$ correlation coefficients of lag 2, and so on. R' is the correlation matrix.

Time-series considerations may provide a theoretical form, Γ, for the correlation matrix R'. Then

$$
V = s_m^2 \Gamma
$$

and weighting can be carried out using Γ^{-1}. The residual variance is s_m^2.

Example   167

7.5 EXAMPLE

7.5.1 Data and analysis

The data are counts of citrus mites made on an orchard on the Central Coast of New South Wales every two weeks for one year. These are part of the data collected by Dr G.A.C. Beattie in a study of citrus pests and their predators. Two orchards were used in the study but only one is considered here. Four plots of four trees were used in each orchard: two plots had been sprayed to control mites two years before these counts were taken; the other two had not been sprayed. There were thus two degrees of freedom for error at each sample time. For simplicity these are assumed to estimate sampling variation only.

Preliminary analyses indicated that there were no differences between sprayed and control plots over the period. The mean incidence over the period was to be described by a regression equation. Periodic regression should be appropriate and several harmonics could be needed because the number of mites follows growth flushes. There are two major growth flushes each year and a number of minor ones depending on climatic conditions.

The mean counts in $\log_e$ over the four plots and the values of the variables needed to fit a periodic regression with three harmonics are given in Table 7.2. The mean variance of the observations over the 26 sampling occasions was 3.536, so that the variance of a plot mean, $s_{\bar{y}}^2$, was 0.8840.

7.5.2 Estimation of R

The average correlation coefficients $r_1, r_2, \ldots, r_{25}$ are plotted against lag in Figure 7.1. This graph is a correlogram and, except for a few large correlations at high lag, which are unreliable, it is similar in shape to the theoretical correlogram of an autoregressive time-series. Hence, it may be possible to model the correlation matrix.

The Markov series is the simplest autoregressive series. In this series the observation at time $t + 1$ depends only on that at time t and on a random deviate. The correlation function is defined for this series by a single parameter ρ because $\rho_n = \rho^n$. This does not appear to be applicable here because $r_1 = 0.68$ and $r_2 = 0.61$, which is not close to 0.68^2. The Yule series is the next simplest autoregressive series. In this series the value of each observation depends on the values of the previous two. The correlation function for the Yule series can be defined by two parameters which are derived from r_1 and r_2. The following expected correlations were obtained assuming a Yule series. They have been used to plot the curve in Figure 7.1 and to produce Γ:

0.688, 0.611, 0.490, 0.409, 0.336, 0.278, 0.229, 0.189, 0.156, 0.129, 0.106, 0.087, 0.072, 0.060, 0.049, 0.041, 0.033, 0.028, 0.023, 0.019, 0.015, 0.013, 0.011, 0.009, 0.007.

Table 7.2 Mean counts (log) of mites on 26 fortnights and variables needed to fit first, second and third harmonics to the means

Fortnight	t	y	t_0	t_{11} $cos(\omega t)$*	t_{12} $sin(\omega t)$	t_{21} $cos(2\omega t)$	t_{22} $sin(2\omega t)$	t_{31} $cos(3\omega t)$	t_{32} $sin(3\omega t)$
1	0	4.912	1	1	0	1	0	1	0
2	1	6.800	1	0.9709	0.2393	0.8855	0.4647	0.7485	0.6631
3	2	5.611	1	0.8855	0.4647	0.5681	0.8230	0.1205	0.9927
4	3	5.800	1	0.7485	0.6631	0.1205	0.9927	−0.5681	0.8230
5	4	5.640	1	0.5681	0.8230	−0.3546	0.9350	−0.9709	0.2393
6	5	5.860	1	0.3546	0.9350	−0.7485	0.6631	−0.8855	−0.4647
7	6	5.200	1	0.1205	0.9927	−0.9709	0.2393	−0.3546	−0.9350
8	7	5.156	1	−0.1205	0.9927	−0.9709	−0.2393	0.3546	−0.9350
9	8	4.018	1	−0.3546	0.9350	−0.7485	−0.6631	0.8855	−0.4647
10	9	3.828	1	−0.5681	0.8230	−0.3546	−0.9350	0.9709	0.2393
11	10	3.397	1	−0.7485	0.6631	0.1205	−0.9927	0.5681	0.8230
12	11	4.369	1	−0.8855	0.4647	0.5681	−0.8230	−0.1205	0.9927
13	12	3.867	1	−0.9709	0.2393	0.8855	−0.4647	−0.7485	0.6631
14	13	4.312	1	−1	0	1	0	−1	0
15	14	4.752	1	−0.9709	−0.2393	0.8855	0.4647	−0.7485	−0.6631
16	15	3.000	1	−0.8855	−0.4647	0.5681	0.8230	−0.1205	−0.9927
17	16	2.266	1	−0.7485	−0.6631	0.1205	0.9927	0.5681	−0.8230
18	17	1.314	1	−0.5681	−0.8230	−0.3546	0.9350	0.9709	−0.2393
19	18	0.000	1	−0.3546	−0.9350	−0.7485	0.6631	0.8855	0.4647
20	19	0.346	1	−0.1205	−0.9927	−0.9709	0.2393	0.3546	0.9350
21	20	1.386	1	0.1205	−0.9927	−0.9709	−0.2393	−0.3546	0.9350
22	21	3.200	1	0.3546	−0.9350	−0.7485	−0.6631	−0.8855	0.4647
23	22	3.884	1	0.5681	−0.8230	−0.3546	−0.9350	−0.9709	−0.2393
24	23	5.200	1	0.7485	−0.6631	0.1205	−0.9927	−0.5681	−0.8230
25	24	7.490	1	0.8855	−0.4647	0.5681	−0.8230	0.1205	−0.9927
26	25	7.193	1	0.9709	−0.2393	0.8855	−0.4647	0.7485	−0.6631

*$\omega = 360/26$.

$$\mathbf{\Gamma} = \begin{bmatrix} 1 & 0.688 & 0.611 & \ldots & & & & 0.009 & 0.007 \\ 0.688 & 1 & 0.688 & & 0.611 & & & & \cdot \\ 0.611 & 0.688 & 1 & & & & & & \cdot \\ & \cdot & & & & & & \cdot & \\ & \cdot & & & & & & & \\ & \cdot & & & & & & & \\ 0.009 & & & & & & & & 0.688 \\ 0.007 & & & & & \ldots & 0.009 & 0.688 & 1 \end{bmatrix}$$

Example   169

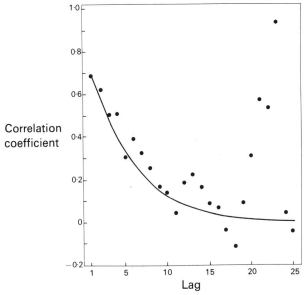

Figure 7.1 Observed correlations of lag 1 to 25 for citrus mites (dots) and correlations predicted by a second-order autoregressive series (solid line).

7.5.3 Results

The values of the mean were regressed on the columns of P in Table 7.2 in ordinary least squares and in weighted least squares using Γ^{-1} as weights. The analysis and the estimates are given in Table 7.3. Although the columns of P are orthogonal to one another, the weighted analysis is non-orthogonal. The sums of squares given for the weighted analysis are those obtained by fitting the terms sequentially, starting with the mean, then adding the first, second and third harmonics in order. The regressions obtained are plotted in Figure 7.2.

7.5.4 Conclusions

There are a number of important points to note from these analyses.

1. The deviations from regression in ordinary least squares had a mean square of 0.3258. This is a serious underestimate of $s_{\bar{y}}^2 = 0.8840$. The deviations from regression always underestimate the variance when the successive observations are positively correlated. Hence, the variances of the estimates must be calculated from Equation (7.2).

2. The estimates of the regression coefficients from the two analyses are very similar. In fact, the fitted regressions are almost indistinguishable over most of the range. This should always be so when V estimates Σ closely.

Table 7.3 Analyses of variance for the regression of mean count on time using ordinary least squares and weighted least squares

Source	DF	Ordinary		Weighted		
		SS	MS	SS	MS	F
Mean	1			85.1750		
First harmonic	2	56.1848		22.7740	11.3870	12.9
Second harmonic	2	21.8607		17.0550	8.5275	9.6
Third harmonic	2	11.3861		9.2540	4.6270	5.2
Deviations from regression (s_y^2)	19	6.1897	0.3258 (0.8840)	18.0380	0.9494 (0.8840)	1.1
Corrected total	25	95.6213				
Total	26			152.2960		

Coefficient	Estimates and standard errors			
b_0	4.18	(0.502)	4.16	(0.501)
b_{11}	1.48	(0.461)	1.50	(0.460)
b_{12}	1.46	(0.552)	1.78	(0.516)
b_{21}	1.24	(0.299)	1.29	(0.297)
b_{22}	-0.37	(0.346)	-0.13	(0.317)
b_{31}	-0.52	(0.221)	-0.44	(0.218)
b_{32}	-0.77	(0.248)	-0.61	(0.228)

Example   171

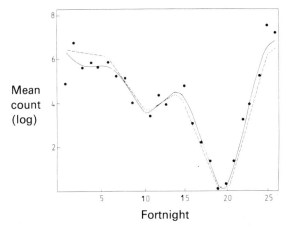

Figure 7.2 Observed counts and periodic regressions fitted by ordinary least squares —— and by weighted least squares ---.

Both methods are then unbiased and therefore estimate the same equation. The weighted estimates may be biased if V is not a close estimate of Σ.

3. The deviations from regression provide a test of lack-of fit in the weighted analysis. However, since there are 19 degrees of freedom for deviations here, it is not a sensitive test. It would probably be wise to fit the fourth and fifth harmonics and test them separately. This would be done by fitting the harmonics in order and testing the significance of the 'extra sums of squares' accounted for by the additional terms (Draper and Smith, 1981, section 2.7).

4. A coefficient of determination, R^2, can be defined for weighted least squares as

$$R_w^2 = \frac{\text{regression sum of squares}}{\text{total sum of squares between fortnights}}.$$

The values for this and a similar one, R_{OLS}^2, for ordinary least squares for the three possible models are given in Table 7.4. R_{OLS}^2 is an overestimate of

Table 7.4

Model	R^2	
	Weighted least squares	*Ordinary least squares*
First harmonic	0.339	0.588
First and second	0.593	0.816
First, second and third	0.731	0.935

R_w^2. This was expected because the deviations underestimate the residual variation. R_{OLS}^2 may be useful for indicating the importance of new terms added to the regression, but it cannot be used to indicate goodness-of-fit.

The estimation of V was simplified in this example by ignoring a between-plot source of variation. In field trials, a between-plot source of variation would be included as well as sampling error; in other experiments a between-individuals source of variation may also be needed to specify the error completely.

7.6 EXPERIMENTS WITH FEW DEGREES OF FREEDOM FOR ERROR

There were only two degrees of freedom for error in the design for the mite example above. The immediate impression was that no worthwhile information would be obtained from a statistical analysis because of the small number of degrees of freedom. However, the means were well determined, each being based on 16 trees. The regression could be fitted to the means directly and the variances of the estimates obtained from

$$\text{var}(\boldsymbol{\beta}) = \boldsymbol{D}\bar{\boldsymbol{S}}_{\bar{y}}\boldsymbol{D}^{\mathrm{T}}. \tag{7.4}$$

Expression (7.4) is similar to expression (7.2) except that $\bar{\boldsymbol{S}}_{\bar{y}}$ is used instead of $\boldsymbol{S}_{\bar{y}}$. In this example the 26 estimates of s^2 would be averaged, as would be the 25 estimates of the correlation coefficient of lag 1, and so on to produce $\bar{\boldsymbol{S}}_{\bar{y}}$. However, it is not clear how many degrees of freedom should be ascribed to the mean variance, s_m^2. If the observations were independent, s_m^2 would be based on 52 degrees of freedom. If this were used, values of a test statistic a little larger than the critical value could not be accepted without reservation. There would also be reservations about the variances obtained from a weighted analysis because the values $(X^{\mathrm{T}}V^{-1}X)^{-1}$ only apply for large samples.

8 Environmental variables

8.1 INTRODUCTION

Trends in treatment comparisons over time were approximated by algebraic or trigonometric polynomials in time or by both in the earlier chapters. The data from the individual time-points were combined and the behaviour of comparisons over time were studied through this approach. However, other variables such as effective rainfall and temperature or soil moisture and soil temperature could be expected to affect the values of the treatment comparisons. Some of the short-term variation in the comparisons may well be due to variation in such variables. The regressions would be more meaningful if these variables were included, particularly if they could replace high-order polynomial terms. It may also be possible to adjust the observed values of the treatment comparisons for unusual environmental conditions.

Environmental effects can be allowed for by extending the within-individual regressions to include environmental variables as well as time-variables. The method will be illustrated for the experiment considered in Chapter 2 for which monthly rainfall was available. The rainfall during the 30 days preceding each weighing was selected as the rainfall to be associated with the live weights at that weighing. Total precipitation was used, although effective rainfall or available water would be used in a serious study of the effects of the environment. Other variables such as soil temperature would also be included. The extra data needed for this approach were not available.

8.2 RAINFALL VARIABLES

A quadratic effect of rain was allowed for initially by using rainfall and its square as variables. The rainfall (r) for the month is given in column 2 of Table 8.1. This was coded as $x_{11} = r - \bar{r}$. The squares of the rainfall were converted to values that were orthogonal to x_{11} as indicated in the table. This step is not essential but can simplify later calculations. The rainfalls for the 30 days preceding the weighings are shown in Figure 8.1, together with the 89-year averages of these periods for the location. The data in Table 2.1 and the variables in Table 2.7 will be used again in this chapter.

Table 8.1   Rainfall (mm) for the 30 days before each of the 27 weighings of Table 2.1 and variables derived from these

Weighing time	rain (r)	x_{11} (r−51)	x_{11}^2	x_{12}' (x_{11}−1299.26)	$(x_{12}'−51.95x_{11})$*	x_{12}
1	59	8	64	−1235.26	−1650.86	−1651
2	35	−16	256	−1043.26	−212.06	−212
3	20	−31	961	−338.26	1272.19	1272
4	99	48	2304	1004.74	−1488.86	−1489
5	100	49	2401	1101.74	−1443.81	−1444
6	108	57	3249	1949.74	−1011.41	−1011
7	100	49	2401	1101.74	−1443.81	−1444
8	38	−13	169	−1130.26	−454.91	−455
9	31	−20	400	−899.26	139.74	140
10	25	−26	676	−623.26	727.44	727
11	42	−9	81	−1218.26	−750.71	−751
12	36	−15	225	−1074.26	−295.01	−295
13	28	−23	529	−770.26	424.59	425
14	166	115	13225	11925.74	5951.49	5951
15	89	38	1444	144.74	−1829.36	−1829
16	24	−27	729	−570.26	832.39	832
17	19	−32	1024	−275.26	1387.14	1387
18	29	−22	484	−815.26	327.64	328
19	23	−28	784	−515.26	939.34	939
20	29	−22	484	−815.26	327.64	328
21	31	−20	400	−899.26	139.74	140
22	27	−24	576	−723.26	523.54	524
23	6	−45	2025	725.74	3063.49	3063
24	55	4	16	−1283.26	−1491.06	−1491
25	63	12	144	−1155.26	−1778.66	−1779
26	49	−2	4	−1295.26	−1191.36	−1191
27	46	−5	25	−1274.26	−1014.51	−1015
Sum	1377	0	35080	0	0	
Mean	51		1299.26			

*$\Sigma x_{11}(x_{11}^2−1299.26)/\Sigma x_{11}^2 = −51.95$.

8.3 SIMPLE REGRESSION OF THE COMPARISONS ON RAINFALL

Regression Equation (8.1) below was fitted to the values of the comparisons as a preliminary step:

$$\hat{c} = b_0 + b_{11}x_{11} + b_{12}x_{12}. \tag{8.1}$$

As usual, two stages were used.

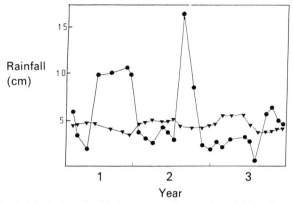

Figure 8.1   Rainfall during the 30 days preceding each weighing ● and the 89-year average rainfall for these periods ▲.

Stage 1

The following new columns were formed from the columns of Table 2.1 and are given in Table 8.2:

$$(b_0) = (4) + (5) + \cdots + (29) + (30),$$
$$(b_{11}) = 8(4) - 16(5) - 31(6) + \cdots + 12(28) - 2(29) - 5(30),$$
$$(b_{12}) = -1651(4) - 212(5) + 1272(6) - \cdots - 1779(28) - 1191(29) - 1015(30);$$

where, as usual, $(\#)$ indicates column $\#$. The multipliers are the values in columns x_{11} and x_{12} of Table 8.1.

The coefficients could be calculated in this simple manner because x_{11} and x_{12} were orthogonal.

Stage 2

The values of b_0, b_{11} and b_{12} were regressed on the x-variables of Table 2.17 to test the significance of the regression on rainfall for each comparison. The analysis of variance given below is that used in section 2.7.2.

Source	DF
Blocks	2
SR	1
P1	1
P1 × SR	1
P2/SR1	1
P2/SR2	1
P3/SR1	1
P3/SR2	1
Error	14
Total	23

Table 8.2   Values of b_0, b_{11} and b_{12} of Equation (8.1) for each plot of Table 2.1

B	P	SR	b_0	$100b_{11}$	$1000b_{12}$
1	1	1	73.9556	−1.3403	1.6998
2	1	1	73.8149	−0.4609	1.9254
3	1	1	68.9556	−2.4552	0.4066
1	2	1	75.2223	0.7028	2.0662
2	2	1	73.0371	3.8718	1.6130
3	2	1	77.5371	0.9564	1.5167
1	1	2	66.9001	0.8433	1.9173
2	1	2	67.4297	4.5323	1.4002
3	1	2	63.0853	0.0480	1.9382
1	2	2	65.3371	1.4111	1.7290
2	2	2	68.6259	4.0947	0.9077
3	2	2	65.5890	3.4094	2.0765
1	3	1	75.4630	4.3758	1.4086
2	3	1	71.9000	3.4861	1.1706
3	3	1	76.5260	3.1326	1.2983
1	4	1	80.7704	5.2021	0.9969
2	4	1	80.6334	3.6771	1.5806
3	4	1	77.8149	4.9353	1.4388
1	3	2	68.2964	3.9111	2.0879
2	3	2	71.3704	6.2954	1.3491
3	3	2	69.5778	5.5277	1.5083
1	4	2	67.6408	8.6423	0.9229
2	4	2	65.1519	9.1622	1.1675
3	4	2	69.3667	4.0003	1.6240

The analyses are given in Table 8.3. The quadratic term was never significant for any comparison and will be ignored in future regressions. The regressions on rainfall were very highly significant for c_3, $c_{5.1}$, $c_{5.2}$ and c_6. The fitted regression equations were:

$$\text{P1: } \hat{c}_3 = 0.93 + 0.0111x_{11}$$
$$(\pm 0.66)(\pm 0.0141)$$

$$\text{P3/SR1: } \hat{c}_{5.1} = 1.72 + 0.0196x_{11}$$
$$(\pm 0.66)(\pm 0.0041)$$

$$\text{P3/SR2: } \hat{c}_{5.2} = 1.20 + 0.0177x_{11} \tag{8.2}$$
$$(\pm 0.66)(\pm 0.0041)$$

$$\text{SR: } \hat{c}_6 = 4.05 - 0.0099x_{11}$$
$$(\pm 0.47)(\pm 0.0029)$$

The regressions are shown in Figure 8.2 where the values of c_3, $c_{5.1}$, $c_{5.2}$ and c_6 are plotted against rainfall. Rainfall did not account for much of the

Table 8.3 Analyses of variance of the values of b_0, b_{11} and b_{12} of Table 8.2

Source	DF	b_0 MS	F	P (%)	$b_{11} \times 10^2$ MS	F	P (%)	$b_{12} \times 10^5$ MS	F	P (%)
Blocks	2	0.86			56.0			929		
Stocking rate	1	394.14	75.0	<0.001	235.6	11.5	0.44	817	0.4	55.1
P1	1	10.47	2.0	18.0	147.0	7.2	1.8	408	0.2	67.2
P1 × SR	1	4.01	0.8	39.7	33.1	1.6	22.5	2408	1.1	31.2
P2/SR1	1	39.17	7.5	1.6	13.3	0.6	43.5	17	0.1	93.2
P2/SR2	1	8.37	1.6	22.8	27.5	1.3	26.6	2400	1.1	31.2
P3/SR1	1	35.31	6.7	2.1	462.2	22.6	0.03	1408	0.6	43.6
P3/SR2	1	17.37	3.3	9.1	374.9	18.3	0.08	1408	0.6	43.6
Error	14	5.26			20.5			2186		

Variable	b_0 Regn coefficient	Standard error	$b_{11} \times 10^2$ Regn coefficient	Standard error	$b_{12} \times 10^5$ Regn coefficient	Standard error
Constant term	71.42	0.47	3.16	0.29	148	9.5
Stocking rate	4.05	0.47	-0.99	0.29	-6	9.5
P1	0.93	0.66	1.11	0.41	6	13.5
P1 × SR	0.58	0.66	0.52	0.41	14	13.5
P2/SR1	-2.56	0.94	-0.47	0.58	-2	19.1
P2/SR2	1.18	0.94	-0.68	0.58	20	19.1
P3/SR1	1.72	0.66	1.96	0.41	-11	13.5
P3/SR2	1.20	0.66	1.77	0.41	-11	13.5

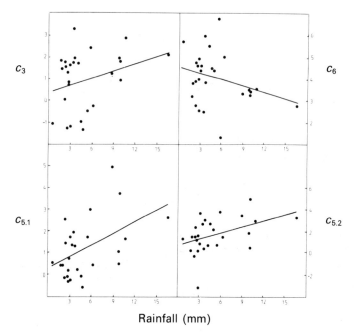

Figure 8.2   Linear regressions on rainfall fitted to the values of c_3, $c_{5.1}$, $c_{5.2}$ and c_6 at 27 weighings.

variation in the comparisons even though the regressions were highly significant.

The regression coefficients obtained describe the simple relations between the values of the comparisons and rainfall. Simple regressions often indicate if the independent variable is likely to be important or not, but not always. Other variables not included in the simple regression will affect the values of the coefficients obtained when they are included in the regression. The regressions in Figure 8.2 indicate that rainfall is not likely to be dominant in this experiment. However, time-variables have already been shown to be important, so the effect of rainfall should be examined in the presence of time-variables. This will be done by adding rainfall to the equations used in Chapter 2 to approximate trends in the values of the comparisons.

The variance of the partial regression coefficient on rainfall and therefore tests of significance are obtained by indirect fitting as usual. However, some equations will also be fitted directly because different equations can often be compared very easily through the values obtained for R^2_{OLS} (section 7.5.4).

The values of $(r-51)$ of Table 8.1 were divided by 100 and added to Table 2.7 to produce Table 8.4. The values of $(r-51)$ were divided by 100 to make rain roughly the same magnitude as the other variables.

8.4 $c_3(P1)$ AND $c_{4.2}$ (P2/SR2)

The following equation was fitted to the values of c_3 and $c_{4.2}$ because the quadratic equations (2.6) and (2.9) had been accepted in section 2.7.1:

$$\hat{c}_3 = b_0 + b_1\xi_1 + b_2\xi_2 + b_{11}x'_{11} \tag{8.3}$$

Stage 1

The live weights for each plot were regressed on the values of ξ_1, ξ_2 and x'_{11} of Table 8.4. The coefficients b_0, b_1, b_2 and b_{11} obtained are given in Table 8.5.

Stage 2

The values of b_0, b_1, b_2 and b_{11} were regressed on the x-variables of Table 2.17. The analysis of variance for b_{11} is given in Table 8.6. Rainfall was now not significant for P1 or P2/SR2 and the quadratic curves of Equation (2.6) and Equation (2.9) were retained.

8.5 $c_6(SR)$

The following equation was fitted to the values of c_6:

$$\hat{c}_6 = b_0 + b_1\xi_1 + b_5\cos d + b_6\sin d + b_8 Y1 \times \sin d + b_{10}Y2 \times \sin d$$
$$+ b_{11}x'_{11}. \tag{8.4}$$

Equation (8.4) extends Equation (2.4) by including rainfall.

8.5.1 Indirect fitting

Stage 1

The live weights for each plot were regressed on the values of ξ_1, $\cos d$, $\sin d$, $Y1 \times \sin d$, $Y2 \times \sin d$ and x'_{11} of Table 8.4. The coefficients obtained are given in Table 8.7.

Stage 2

The values of b_0, b_1, b_5, b_6, b_8, b_{10} and b_{11} were regressed on the x-variables of Table 2.17. The analysis of variance for b_{11} is given in Table 8.8. Rainfall was now not significant for SR.

8.5.2 Direct fitting

The values of c_6 given in Table 2.6 were regressed directly on the values of ξ_1, $\cos d$, $\sin d$, $Y1 \times \sin d$, $Y2 \times \sin d$ and x'_{11} of Table 8.4 and on subsets of these variables. The values of R^2_{OLS} obtained are given in Table 8.9. R^2_{OLS}

Table 8.4 Live weights for plots 1 and 24 and regression variables

t	d	Live weights plot 1	...	24	ξ_0	ξ_1	ξ_2	ξ_3	ξ_4
0	0	657	...	734	1	−0.3378	0.3954	−0.3699	0.3488
29	29	635		640	1	−0.3210	0.3351	−0.2403	0.1191
85	85	679		639	1	−0.2887	0.2277	−0.0396	−0.1681
141	141	718		696	1	−0.2564	0.1321	0.1017	−0.2928
212	212	801		766	1	−0.2154	0.0278	0.2067	−0.2898
287	287	678		626	1	−0.1721	−0.0618	0.2423	−0.1736
302	302	726		764	1	−0.1635	−0.0772	0.2416	−0.1434
358	358	724		686	1	−0.1358	−0.1208	0.2247	−0.0422
386	21	714		681	1	−0.1156	−0.1471	0.2003	0.0296
426	61	734		677	1	−0.0925	−0.1715	0.1622	0.1027
478	113	765		683	1	−0.0636	−0.1939	0.1021	0.1735
507	142	800		722	1	−0.0463	−0.2023	0.0640	0.1997
546	181	856		801	1	−0.0238	−0.2086	0.0103	0.2167
581	216	798		805	1	−0.0030	−0.2093	−0.0397	0.2132
636	271	810		822	1	0.0287	−0.2010	−0.1128	0.1732
686	321	850		844	1	0.0587	−0.1827	−0.1726	0.1016
728	363	788		743	1	0.0830	−0.1605	−0.2105	0.0260
747	17	808		724	1	0.0939	−0.1483	−0.2236	−0.0114
776	46	788		695	1	0.1107	−0.1270	−0.2381	−0.0700
824	94	722		617	1	0.1384	−0.0849	−0.2448	−0.1635
923	193	808		736	1	0.1955	0.0294	−0.1733	−0.2729
936	206	804		734	1	0.2030	0.0471	−0.1538	−0.2716
972	242	766		695	1	0.2238	0.0996	−0.0858	−0.2392
1021	291	637		564	1	0.2520	0.1788	0.0426	−0.1107
1049	319	647		564	1	0.2682	0.2281	0.1362	0.0173
1077	347	632		551	1	0.2844	0.2804	0.2456	0.1931
1095	365	623		538	1	0.2947	0.3155	0.3247	0.3345

was not increased by adding x'_{11} to the regression equation and was decreased when x'_{11} was substituted for $Y1 \times \sin d$ or $Y2 \times \sin d$. Equation (2.5) was retained.

8.6 $c_{5.1}$(P3/SR1)

Adding rainfall to Equation (2.3) provided the following equation to be fitted to the values of $c_{5.1}$:

$$\hat{c}_{5.1} = b_0 + b_1\xi_1 + b_2\xi_2 + b_3\xi_3 + b_4\xi_4 + b_5\cos d + b_6\sin d$$
$$+ b_7 Y1 \times \cos d + b_8 Y1 \times \sin d + b_9 Y2 \times \cos d$$
$$+ b_{10} Y2 \times \sin d + b_{11} x'_{11}. \tag{8.5}$$

cos d	sin d	$Y1 \times \cos$	$Y1 \times \sin$	$Y2 \times \cos$	$Y2 \times \sin$	$rain \times 10^{-2}$ (x_{11}^1)
1	0	1	0	1	0	0.08
0.8781	0.4785	0.8781	0.4785	0.8781	0.4785	−0.16
0.1081	0.9941	0.1081	0.9941	0.1081	0.9941	−0.31
−0.7547	0.6561	0.7547	0.6561	−0.7547	0.6561	0.48
−0.8747	−0.4847	0.8747	−0.4847	−0.8747	−0.4847	0.49
0.2237	−0.9747	0.2237	−0.9747	0.2237	−0.9747	0.57
0.4650	−0.8853	0.4650	−0.8853	0.4650	−0.8853	0.49
0.9924	−0.1233	0.9924	−0.1233	0.9924	−0.1233	−0.13
0.9366	0.3505	−0.9366	−0.3505	0.9366	0.3505	−0.20
0.5007	0.8656	−0.5007	−0.8656	0.5007	0.8656	−0.26
−0.3618	0.9322	0.3618	−0.9322	−0.3618	0.9322	−0.09
−0.7638	0.6454	0.7638	−0.6454	−0.7638	0.6454	−0.15
−0.9995	0.0306	0.9995	−0.0306	−0.9995	0.0306	−0.23
−0.8410	−0.5410	0.8410	0.5410	−0.8410	−0.5410	1.15
−0.0529	−0.9986	0.0529	0.9986	−0.0529	−0.9986	0.38
0.7225	−0.6914	−0.7225	0.6914	0.7225	−0.6914	−0.27
0.9992	−0.0408	0.9992	0.0408	0.9992	−0.0408	−0.32
0.9593	0.2822	0	0	−1.9187	−0.5645	−0.22
0.7073	0.7069	0	0	−1.4147	−1.4138	−0.28
−0.0401	0.9992	0	0	0.0803	−1.9984	−0.22
−0.9851	−0.1718	0	0	1.9703	0.3437	−0.20
−0.9225	−0.3861	0	0	1.8449	0.7721	−0.24
−0.5270	−0.8499	0	0	1.0540	1.6998	−0.45
0.2841	−0.9588	0	0	−0.5681	1.9176	−0.04
0.6960	−0.7181	0	0	−1.3920	1.4361	0.12
0.9495	−0.3139	0	0	−1.8989	0.6277	−0.02
1	−0.009	0	0	−2	0.0191	−0.05

8.6.1 Indirect fitting

Stage 1

The live weights for each plot were regressed on the values of $\xi_1, \ldots, x_{11}'$ in Table 8.4. The coefficients $b_0, \ldots, b_{11}$ obtained are given in Table 8.10.

Stage 2

The values of $b_0, \ldots, b_{11}$ were regressed on the x-variables of Table 2.17. The analysis of variance for b_{11} is given in Table 8.11 and the fitted values of all the coefficients in Equation (8.5) are given in Table 8.12. Rainfall was still highly significant, but $Y2 \times \sin d$ was no longer significant at $P = 0.05$.

Table 8.5   Values of b_0, b_1, b_2 and b_{11} of Equation (8.3) for each plot of Table 2.1

B P SR	b_0 (constant)	b_1 (linear)	b_2 (quadratic)	b_{11} (rainfall)
1 1 1	73.9506	1.2212	−28.2100	−1.6980
2 1 1	73.8152	7.1057	−25.8134	0.0625
3 1 1	68.9428	−9.7616	−23.2965	−4.2788
1 2 1	75.2173	−10.8487	−38.4103	−1.6251
2 2 1	73.0395	−16.0691	−36.3224	0.9006
3 2 1	77.5354	−7.1321	−26.9963	−0.5284
1 1 2	66.9018	−0.4226	−24.0234	0.5791
2 1 2	67.4392	−6.9461	−19.1402	3.2528
3 1 2	63.0785	−13.4686	−30.3337	−2.2182
1 2 2	65.3338	−15.2136	−28.9896	−1.0454
2 2 2	68.6283	−19.9904	−22.2301	0.8994
3 2 2	66.3329	−14.3660	−30.0782	1.1431
1 3 1	75.4693	−12.0916	−26.4955	2.1924
2 3 1	71.9046	−10.7584	−21.9792	1.5903
3 3 1	76.5312	−6.5754	−24.5801	1.7929
1 4 1	80.7801	−10.6426	−23.5327	3.3351
2 4 1	80.6394	−7.4409	−31.7197	2.0751
3 4 1	77.8233	−10.8245	−28.3975	2.9006
1 3 2	68.6739	−3.8596	−35.6265	2.4109
2 3 2	71.3809	−15.2703	−29.4526	3.6138
3 3 2	69.5845	−18.1328	−32.3968	2.3533
1 4 2	67.6549	−24.4501	−26.1369	4.8928
2 4 2	65.1635	−19.9377	−21.8247	4.0230
3 4 2	69.3720	−11.9593	−25.7387	1.8412

Table 8.6   Analysis of variance of the values of b_{11} in Table 8.5

Source	DF	MS	F	P(%)
Blocks	2	5.640		
Stocking rate	1	9.408		
P1	1	1.363	0.73	40.8
P1/SR	1	2.321		
P2/SR1	1	1.247		
P2/SR2	1	0.943	0.50	49.0
P3/SR1	1	36.938		
P3/SR2	1	22.754		
Error	14	1.877		

Table 8.7 Values of b_0, b_1, b_5, b_6, b_8, b_{10} and b_{11} of Equation (8.4) for each plot of Table 2.1

B P SR	b_0 (constant)	b_1 (linear)	b_5 (cos)	b_6 (sin)	b_8 ($Y1 \times sin$)	b_{10} ($Y2 \times sin$)	b_{11} (rainfall)
1 1 1	73.3482	−2.6393	−6.6870	−2.5873	0.4300	−4.0542	−9.8133
2 1 1	75.2572	−0.3494	−7.7933	−4.6379	0.4744	−3.7514	−10.5118
3 1 1	70.0392	−19.6662	−7.3640	−6.0472	−1.0442	−2.3066	−15.0494
1 2 1	76.6311	−17.0493	−7.2500	−4.1093	0.4515	−4.0397	−11.2075
2 2 1	74.6536	−23.3431	−8.0906	−5.1647	−1.3506	−5.6390	−11.2654
3 2 1	79.2635	−13.7533	−8.3192	−4.3458	1.0318	−5.2900	−12.1192
1 1 2	68.4746	−10.4632	−8.1005	−6.3540	2.2084	−4.7153	−12.5424
2 1 2	68.5996	−16.8447	−5.4513	−6.6821	2.4338	−4.8379	−8.9294
3 1 2	64.7349	−15.9191	−6.7840	−2.0897	0.8473	−5.9084	−11.0929
1 2 2	66.5352	−20.7676	−5.4259	−4.0816	0.3844	−4.7310	−10.3176
2 2 2	69.6308	−26.6778	−5.0638	−4.5183	0.6241	−3.5742	−8.0206
3 2 2	67.7244	−21.4236	−6.6415	−4.8725	0.8485	−4.9174	−9.6705
1 3 1	76.6784	−21.6341	−6.3437	−6.3571	0.9116	−4.3169	−9.5861
2 3 1	73.1507	−18.7107	−7.3083	−4.9115	−0.4426	−3.0689	−8.6165
3 3 1	77.5366	−13.6320	−5.6927	−4.6103	−0.3074	−3.0226	−7.1139
1 4 1	81.7240	−20.8252	−5.2453	−7.0081	−0.4972	−4.1867	−8.4295
2 4 1	82.0044	−16.4800	−6.7455	−6.2952	0.2996	−5.3018	−10.3622
3 4 1	79.0793	−18.2685	−7.0896	−4.7896	−0.6208	−3.4632	−7.1375
1 3 2	70.4612	−12.2540	−8.2303	−5.6680	2.5795	−6.0500	−10.6798
2 3 2	72.5526	−26.0003	−5.6268	−7.5148	1.2639	−5.3869	−9.5087
3 3 2	70.9191	−28.6431	−6.3236	−7.4471	0.8644	−5.9515	−11.3779
1 4 2	68.6140	−33.4585	−4.7213	−6.5316	−0.5609	−4.8810	−6.5062
2 4 2	66.2259	−31.2044	−4.8031	−7.7138	3.1990	−5.0879	−8.8881
3 4 2	70.5891	−20.5618	−5.8680	−5.7046	2.4661	−4.3244	−9.0983

Table 8.8 Analysis of variance of the values of b_{11} in Table 8.7

Source	DF	MS	F	P(%)
Blocks	2	1.347		
Stocking rate	1	0.874	0.32	57.8
P1	1	2.375		
P1 × SR	1	1.187		
P2/SR1	1	0.063		
P2/SR2	1	8.340		
P3/SR1	1	29.206		
P3/SR2	1	1.698		
Error	14	2.694		

Table 8.9

ξ_1	cos d	sin d	Y1 × sin d	Y2 × sin d	x_{11}	R^2_{OLS}
		Variables in the regression				
+	+	+	+	+	+	0.719
+	+	+	+	+		0.717
+	+	+	+		+	0.585
+	+	+		+	+	0.606
+	+	+			+	0.468

The values predicted for $c_{5.1}$ using Equations (2.10) and (8.5) are given in Table 8.13 and are plotted in Figure 8.3.

8.6.2 Direct fitting

The values of $c_{5.1}$ given in Table 2.18 were regressed directly on the values of ξ_1, ξ_2, ξ_3, ξ_4, cos d, sin d, Y1 × cos d, Y1 × sin d, Y2 × cos d, Y2 × sin d and

Table 8.10   Values of the coefficients of Equation (8.5) for each plot

B P SR	b_0 (constant)	b_1 (linear)	b_2 (quad.)	b_3 (cubic)	b_4 (quartic)	b_5 (cos)
1 1 1	74.7987	−7.5356	−22.7113	−9.6149	4.6601	−5.1449
2 1 1	74.8031	−2.6161	−19.2299	−10.3493	5.4987	−6.3450
3 1 1	69.8486	−30.2743	−17.4983	−6.5196	8.6417	−7.4759
1 2 1	75.8164	−24.4338	−37.2961	−11.8406	12.4990	−5.2720
2 2 1	74.0780	−28.2056	−27.1451	−10.1687	5.8448	−6.2010
3 2 1	78.5199	−23.0010	−17.7492	−20.9920	1.6627	−7.0684
1 1 2	68.3777	−13.3203	−13.6839	−9.5912	9.3357	−7.6496
2 1 2	68.0295	−21.3473	−13.5996	−15.9077	0.5870	−4.1336
3 1 2	64.5770	−20.1909	−17.8592	−8.0238	11.6516	−6.3774
1 2 2	66.2069	−20.1013	−22.7189	−4.6900	11.1316	−4.0587
2 2 2	69.3847	−33.1838	−16.6168	−10.3364	9.1928	−4.7134
3 2 2	67.4777	−27.3002	−21.6554	−10.8256	13.8979	−6.1710
1 3 1	76.3882	−24.4498	−20.2622	−8.2770	10.3002	−5.3986
2 3 1	73.5241	−15.2076	−10.9218	4.5976	15.4881	−7.5036
3 3 1	77.4515	−12.1371	−19.3395	0.6920	12.1125	−4.8411
1 4 1	81.6006	−19.0609	−16.7529	−0.2808	7.0897	−4.0814
2 4 1	81.6878	−17.2307	−23.8443	−7.1234	10.4633	−5.2663
3 4 1	79.2892	−17.1228	−19.3234	4.5658	17.6606	−6.9505
1 3 2	70.1079	−17.7281	−25.930	−15.7053	21.9585	−7.9320
2 3 2	72.3002	−28.4464	−22.5216	−10.1142	12.9822	−4.6100
3 3 2	70.4778	−29.8469	−24.9053	−13.4575	10.8050	−4.6020
1 4 2	68.4276	−35.3383	−18.9140	−0.0384	10.1247	−4.0838
2 4 2	66.2711	−32.9490	−13.9855	−9.9935	14.6195	−4.6970
3 4 2	70.2764	−22.2754	−20.6480	−12.9293	13.1931	−4.8791

x'_{11} of Table 8.4 and on subsets of these variables. The values of R^2_{OLS} obtained for the regressions are given in Table 8.14.

When x'_{11} (rain) was included in the regression, nothing worthwhile was gained by including $\sin d$, $Y1 \times \cos d$, $Y1 \times \sin d$ or $Y2 \times \sin d$. The reduced equation was

$$\hat{c}_{5.1} = 1.74 + 0.011\xi_1 + 1.92\xi_2 + 3.27\xi_3 + 2.92\xi_4 + 0.40 \cos d$$
$$+ 1.08 \ Y2 \times \cos d + 2.47 \ x'_{11}. \qquad (8.6)$$

The values of $c_{5.1}$ predicted by Equation (8.6) are given in Table 8.13.

8.7 $c_{5.2}(\text{P3/SR2})$

The following equation was fitted to the values of $c_{5.2}$:

$$\hat{c}_{5.2} = b_0 + b_5\cos d + b_6\sin d + b_7 Y1 \times \cos d + b_9 Y2 \times \cos d + b_{11}x'_{11}. \ (8.7)$$

b_6 (sin)	b_7 ($Y1 \times cos$)	b_8 ($Y1 \times sin$)	b_9 ($Y2 \times cos$)	b_{10} ($Y2 \times sin$)	b_{11} (rain)
-3.7355	-2.0444	0.7469	-0.9668	-1.0456	-7.6996
-5.1813	-1.7748	0.3600	-0.2961	-1.0217	-7.1659
-8.6666	-4.2123	-0.8767	-2.0237	-0.2480	-17.1252
-6.0642	-2.2049	1.6165	-1.2449	0.5480	-8.9526
-6.2088	-2.8344	-0.7594	-0.7543	-2.1071	-8.2131
-6.7371	-2.7366	-1.0526	-2.6766	-2.0789	-10.5978
-7.0426	-3.5807	1.5742	-0.0547	-2.6810	-9.8383
-7.8194	-1.7460	0.9098	-1.3968	-2.3204	-6.2584
-3.2036	-3.3367	0.8325	-0.2967	-3.6019	-9.3725
-3.9819	-1.0256	1.5205	0.8876	-2.0380	-6.5271
-6.1915	-3.4686	0.1164	-1.1056	-1.2604	-7.1899
-6.4469	-3.8443	0.6512	-0.6380	-2.0619	-7.8864
-7.0924	-2.5263	1.1407	-0.0506	-1.6885	-6.8690
-4.0411	-2.2878	1.0689	2.2726	-2.2856	-6.0073
-4.2504	-1.2360	1.4056	1.3522	-1.0669	-4.2444
-6.4228	-1.3668	0.8804	1.1809	-2.3639	-5.1035
-6.4112	-2.5019	1.0730	0.6182	-2.3125	-6.1475
-4.5068	-2.5968	1.6094	1.8348	-1.8441	-5.0306
-7.5269	-3.1983	1.8672	-0.4271	-2.5013	-7.9908
-8.1420	-3.4302	1.3234	0.2655	-2.3935	-5.6933
-7.7332	-2.9616	0.6414	0.3600	-2.4080	-5.8783
-7.0550	-1.1196	0.9964	0.2663	-2.9808	-5.6172
-8.2292	-4.1217	2.4357	0.5500	-3.0291	-5.4717
-6.2853	-2.7554	1.9030	0.2519	-1.3352	-4.7956

Table 8.11 Analysis of variance of the values of b_{11} in Table 8.9

Source	DF	MS	F	P(%)
Blocks	2	5.107		
Stocking rate	1	4.714		
P1	1	5.458		
P1 × SR	1	0.011		
P2/SR1	1	0.117		
P2/SR2	1	2.254		
P3/SR1	1	57.869	11.2	0.47
P3/SR2	1	11.263		
Error	14	5.146		

Table 8.12 Fitted values for coefficients of Equations (8.5) and (2.10) and P-values for these from the analyses of $b_0, \ldots, b_{11}$

Coefficient	Equation (8.5)		Equation (2.10)	
	Fitted value	P(%)	Fitted value	P(%)
b_0	1.8400	0.91	1.893	0.82
b_1	0.905	68.7	−1.030	59.4
b_2	2.599	6.7	3.120	2.8
b_3	5.305	0.37	6.238	0.13
b_4	2.859	4.5	3.355	1.9
b_5	0.289	48.2	−0.127	72.4
b_6	0.322	50.9	−0.327	41.7
b_7	0.274	35.0	0.221	44.0
b_8	0.595	1.3	0.680	0.80
b_9	1.264	0.02	1.050	0.04
b_{10}	−0.467	6.0	−0.757	0.71
b_{11}	2.20	0.47	−	−

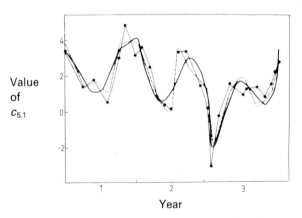

Value of $c_{5.1}$

Year

Figure 8.3 Observed values of $c_{5.1}$ and those predicted by Equation (2.10) —— and by Equation (8.5) ---.

Table 8.13 Original estimates of $c_{5.1}$ for 27 weighings and those predicted by Equations (2.10), (8.5) and (8.6)

Weighing	Original estimates	Values of $c_{5.1}$ Values predicted by Equation		
		(2.10)	(8.5)	(8.6)
1	3.84	3.48	3.60	3.98
2	2.72	2.98	2.95	2.85
3	1.48	1.83	1.46	0.96
4	1.82	1.09	1.62	1.54
5	0.57	1.71	1.25	1.54
6	3.03	3.46	3.53	3.65
7	4.98	3.74	3.89	3.86
8	3.07	4.10	3.95	3.27
9	3.62	2.94	3.01	3.09
10	2.42	1.63	1.90	2.34
11	0.72	0.67	0.97	1.45
12	0.42	0.70	0.40	0.64
13	0.21	1.30	0.15	−0.04
14	3.38	2.08	3.54	3.43
15	3.46	2.84	2.75	2.34
16	1.45	2.25	1.64	1.52
17	0.23	0.85	1.05	1.51
18	−3.14	−1.90	−1.98	−1.53
19	−0.26	−1.07	−1.16	−1.42
20	1.59	0.68	0.87	−0.17
21	0.94	1.78	1.94	1.67
22	1.23	1.56	1.59	1.56
23	1.38	0.80	0.24	0.77
24	0.78	0.32	0.63	1.31
25	1.48	0.83	1.20	1.75
26	2.19	2.17	2.10	1.94
27	2.70	3.48	3.25	2.51

Table 8.14

ξ_1	ξ_2	ξ_3	ξ_4	$\cos d$	$\sin d$	$Y1 \times \cos d$	$Y1 \times \sin d$	$Y2 \times \cos d$	$Y2 \times \sin d$	x'_{11}	R^2_{OLS}
+	+	+	+	+	+	+	+	+	+	+	0.85
+	+	+	+	+	+	+	+	+	+		0.77
+	+	+	+	+	+	+	+	+		+	0.81
+	+	+	+	+	+	+	+		+	+	0.47
+	+	+	+	+	+	+		+	+	+	0.83
+	+	+	+	+	+		+	+	+	+	0.84
+	+	+	+	+				+		+	0.78
+	+	+	+	+				+			0.56

Variables in the regression

8.7.1 Indirect fitting

Stage 1

The live weights for each plot were regressed on the values of $\cos d$, $\sin d$, $Y1 \times \cos d$, $Y2 \times \cos d$ and x'_{11} of Table 8.4. The values obtained for the coefficients are given in Table 8.15.

Stage 2

The values of b_0, b_5, b_6, b_7, b_9 and b_{11} were regressed on the x-variables of Table 2.17. The analysis of b_{11} is given in Table 8.16. Rainfall was still highly significant. The fitted equation was:

$$\hat{c}_{5.2} = 1.107 + 0.466 \cos d - 0.261 \sin d + 0.305 \ Y1 \times \cos d$$
$$(\pm 0.636) \ (\pm 0.330) \quad (\pm 0.393) \qquad (\pm 0.174)$$
$$+ 0.471 \ Y2 \times \cos d + 2.03 \ x'_{11}.$$
$$(\pm 0.176) \qquad\qquad (\pm 0.0073)$$

(8.8)

The values predicted from this equation and from Equation (2.11) are given in Table 8.17 and are plotted in Figure 8.4.

8.7.2 Direct fitting

The values of $c_{5.2}$ given in Table 2.18 were regressed directly on the values of $\cos d$, $\sin d$, $Y1 \times \cos d$, $Y2 \times \cos d$ and x'_{11} of Table 8.4 and on subsets of these variables. The values of R^2_{OLS} obtained for these regressions are given in Table 8.18.

When x'_{11} was included little was gained by including $\sin d$ and $Y1 \times \cos d$. The reduced equation was:

$$\hat{c}_{5.2} = 1.14 + 0.48 \cos d + 0.50 \ Y2 \times \cos d + 2.32 \ x'_{11}.$$

(8.9)

The values predicted by this equation are given in Table 8.17.

8.8 DISCUSSION

The analyses carried out in this chapter quantified the effect of rainfall on the treatment comparisons. No effects of rainfall were established for $c_3(\text{P1})$, $c_4(\text{P2})$ or $c_6(\text{SR})$, but $c_5(\text{P3})$ was markedly affected by rainfall. Thus no effect of rainfall was established on the differences between pastures that contained phalaris and those that did not (comparisons P1 and P2). However, the differences between pastures with lucerne and those without (P3) depended markedly on rainfall at both stocking rates.

The rainfall over the period of the experiment differed noticeably from the 89-year average (Figure 8.1); there were periods in which rainfall was well

Table 8.15   Values of b_0, b_5, b_6, b_7, b_9 and b_{11} of Equation (8.7) for each plot of Table 2.1

B P SR	b_0 (constant)	b_5 (cos)	b_6 (sin)	b_7 ($Y1 \times cos$)	b_9 ($Y2 \times cos$)	$100b_{11}$ (rainfall)
1 1 1	74.9088	−5.5243	−1.0084	−1.6409	0.2060	−4.2067
2 1 1	74.8338	−6.7008	−3.4120	−1.2692	0.4193	−5.6798
3 1 1	70.0156	−6.0615	−2.0664	−2.1342	0.7141	−5.8800
1 2 1	76.2005	−5.3571	−0.0132	−0.5260	1.1927	−0.7803
2 2 1	74.2129	−5.3021	0.8517	−1.5022	2.2953	4.1627
3 2 1	78.6502	−6.4006	−0.6815	−1.2630	0.5401	−1.8981
1 1 2	67.9933	−6.2165	−2.9860	−1.9653	1.3939	−2.6572
2 1 2	68.0091	−3.2932	−2.3811	−0.6374	1.4272	2.5348
3 1 2	64.1132	−4.3160	2.5335	−1.3517	1.6654	1.7459
1 2 2	66.0148	−2.7765	1.3090	0.5172	2.6619	3.3648
2 2 2	69.3757	−2.7924	1.2894	−1.0243	2.0735	5.7089
3 2 2	67.2649	−4.2053	0.4307	−1.2392	1.9314	3.9163
1 3 1	76.2704	−3.9413	−1.0674	−0.6102	2.2711	3.6827
2 3 1	72.9637	−5.1302	−0.0926	−0.2913	2.8871	3.2478
3 3 1	77.2686	−3.8987	−0.9303	0.0042	2.0543	2.2452
1 4 1	81.3877	−2.7328	−1.6169	−0.3583	2.8007	5.0978
2 4 1	81.5031	−4.2239	−1.4638	−1.0927	2.3227	2.7348
3 4 1	78.8484	−4.9196	−0.0463	−0.5927	2.5718	4.7190
1 3 2	69.6123	−5.8993	−1.8663	0.0908	1.2401	0.4062
2 3 2	72.0619	−2.6316	−1.0055	−1.0038	2.9576	6.9814
3 3 2	70.3715	−2.9414	−0.1885	−0.6868	3.4637	6.9673
1 4 2	68.1424	−1.5018	1.0427	0.8852	3.1964	11.3644
2 4 2	65.7529	−1.5962	−0.3489	−0.9719	3.4870	9.2782
3 4 2	70.1132	−3.4366	−0.5652	−0.3536	2.4558	4.0045

Table 8.16   Analysis of variance of b_{11} in Table 8.15

Source	DF	MS	F	P(%)
Blocks	2	6.965		
Stocking rate	1	88.821		
P1	1	68.246		
P1 × SR	1	2.885		
P2/SR1	1	1.899		
P2/SR2	1	17.655		
P3/SR1	1	108.057		
P3/SR2	1	49.567	7.8	1.4
Error	14	6.361		

Table 8.17 Original estimates of $c_{5.2}$ for 27 weighings and those predicted by Equations (2.11), (8.8) and (8.9)

| Weighing | | Values of $c_{5.2}$ | | |
| | Original estimates | Values predicted by Equation | | |
		(2.11)	(8.8)	(8.9)
1	3.02	2.00	2.52	2.31
2	1.38	1.54	1.76	1.63
3	0.48	0.46	0.36	0.53
4	0.50	−0.07	0.99	1.51
5	1.04	0.66	1.14	1.42
6	1.65	2.02	2.79	2.68
7	3.74	2.17	2.91	2.73
8	1.32	2.09	2.10	1.81
9	1.76	0.88	1.21	1.59
10	1.49	0.48	0.68	1.03
11	0.29	0.39	0.46	0.58
12	1.96	0.59	0.15	0.04
13	0.22	1.03	−0.02	−0.37
14	2.60	1.56	3.07	2.98
15	4.91	1.83	2.09	1.97
16	2.56	1.64	1.17	1.22
17	0.48	1.17	1.09	1.38
18	−0.28	0.57	0.11	0.13
19	−0.10	0.34	0.01	0.12
20	0.78	0.37	0.43	0.65
21	−0.21	1.55	1.21	1.19
22	−0.08	1.69	1.15	1.06
23	0.54	1.90	0.63	0.37
24	1.71	1.71	1.28	0.90
25	0.48	1.39	1.18	1.06
26	−0.58	1.01	0.68	0.60
27	−0.02	0.77	0.52	0.50

above average in the first and second years; the rainfall in the third year was generally below average. Equations (8.5) or (8.6) and (8.8) or (8.9) allow the values of $c_{5.1}$ and $c_{5.2}$ to be adjusted for the difference between the rainfall during the experiment and the 89-year average. In this way the results of the experiment can be generalized more reliably. The adjustments are made using $b(R_t - r_t)$, where b is the partial regression coefficient of c_5 on rainfall and R_t and r_t are the average and actual rainfall at the tth weighing. Adjusted values for $c_{5.1}$ and $c_{5.2}$ are calculated in Table 8.19 using the partial regression coefficients of Equations (8.5) and (8.8).

The partial regression coefficients of $c_{5.1}$ and $c_{5.2}$ on rainfall were 0.0221 ± 0.0066 and 0.0203 ± 0.0073. These agree closely. The corresponding simple regression coefficients of Equation (8.2) were 0.0196 ± 0.0041 and 0.0177 ± 0.0041. These are close to the partial regression coefficients; how-

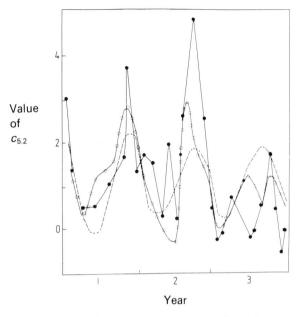

Figure 8.4   Observed values of $c_{5.2}$ ——— and those predicted from Equation (2.11) -- and by Equation (8.8) ⊖.

Table 8.18

Variables in the regression					R^2_{OLS}
$\cos d$	$\sin d$	$Y1 \times \cos d$	$Y2 \times \cos d$	x_{11}	
+	+	+	+	+	0.46
+	+	+	+		0.25
+	+	+		+	0.32
+	+		+	+	0.44
+	+			+	0.30
+			+	+	0.43

ever, the importance of rainfall is shown much more clearly in Figures 8.3 and 8.4 than in Figure 8.2. Figures 8.3 and 8.4 are based on multiple regressions of $c_{5.1}$ and $c_{5.2}$ on time and rainfall; Figure 8.2 is based on their simple regressions on rainfall.

Although the importance of rainfall was established through multiple regression in the usual way, this did not provide much insight into how rainfall produced its effect in the regression. This important aspect should be investigated. The first step is to regress rainfall on the time-variables already present in the equation. If rainfall were completely expressible in terms of the time-variables of Equation (8.5), for example, nothing could be gained by including rainfall in the regression. It would, however, substitute

Table 8.19 Calculation of values of $c_{5.1}$ and $c_{5.2}$ adjusted for rainfall

Weighing	Rainfall**		Adjustments*			Values of $c_{5.1}$		Values of $c_{5.2}$	
	R_t	r_t	$R_t - r_t$	$c_{5.1}$	$c_{5.2}$	Original estimates	Adjusted estimates	Original estimates	Adjusted estimates
1	41	59	−18	−0.40	−0.36	3.84	3.44	3.02	2.66
2	43	35	8	0.18	0.16	2.72	2.90	1.38	1.54
3	48	20	28	0.62	0.57	1.48	2.10	0.48	1.05
4	47	99	−52	−1.15	−1.06	1.82	0.67	0.50	−0.56
5	40	100	−60	−1.33	−1.22	0.57	−0.76	1.04	−0.18
6	38	108	−70	−1.55	−1.42	3.03	1.48	1.65	0.23
7	37	100	−63	−1.39	−1.28	4.98	3.59	3.74	2.46
8	41	38	3	0.07	0.06	3.07	3.14	1.32	1.38
9	42	31	11	0.24	0.22	3.62	3.86	1.76	1.98
10	50	25	25	0.55	0.51	2.42	2.97	1.49	2.00
11	48	42	6	0.13	0.12	0.72	0.85	0.29	0.41
12	47	36	11	0.24	0.22	0.42	0.66	1.96	2.18
13	52	28	24	0.53	0.49	0.21	0.74	0.22	0.71
14	38	166	−128	−2.83	−2.60	3.38	0.55	2.60	0
15	39	89	−50	−1.10	−1.02	3.46	2.36	4.91	3.89
16	39	24	15	0.33	0.30	1.45	1.78	2.56	2.86
17	41	19	22	0.49	0.45	0.23	0.72	0.48	0.93
18	42	29	13	0.29	0.26	−3.14	−2.85	−0.28	−0.02
19	47	23	24	0.53	0.49	−0.26	0.27	−0.10	0.39
20	47	29	18	0.40	0.37	1.59	1.99	0.78	1.15
21	48	31	17	0.38	0.35	0.94	1.32	−0.21	0.14
22	42	27	15	0.33	0.30	1.23	1.56	−0.08	0.22
23	39	6	33	0.73	0.67	1.38	2.11	0.54	1.24
24	38	55	−17	−0.38	−0.35	0.78	0.40	1.71	1.36
25	38	63	−25	−0.55	−0.51	1.48	0.93	0.48	−0.03
26	40	49	−9	−0.20	−0.18	2.19	1.99	−0.58	−0.76
27	41	46	−5	−0.11	−0.10	2.70	2.59	−0.02	−0.12

*Adjustment = 0.0220 $(R_t - r_t)$ for $c_{5.1}$ and 0.0203 $(R_t - r_t)$ for $c_{5.2}$.
**R_t – 89-year average rainfall. r_t – observed rainfall.

for some of the time-variables. Environmental variables will never be completely expressible as a function of time-variables. The strength of the relation between the environmental variable and the time-variables will determine the importance of the additional variable in the regression. These aspects will be illustrated for the regressions of $c_{5.1}$ and $c_{5.2}$ on time and rainfall.

The values of rainfall (r) in Table 8.1 were regressed on the values of (a) ξ_1, ξ_2, ξ_3, ξ_4, $\cos d$, $\sin d$, $Y1 \times \cos d$, $Y1 \times \sin d$, $Y2 \times \cos d$ and $Y2 \times \sin d$ of Table 8.4, and (b) $\cos d$, $\sin d$, $Y1 \times \cos d$ and $Y2 \times \cos d$. These sets of time-variables appear in Equations (8.5) and (8.7). The equations obtained were:

$$\hat{r} = 53.5 - 87.9\xi_1 + 22.8\xi_2 + 40.6\xi_3 + 20.4\xi_4 - 18.5 \cos d$$
$$- 30.3 \sin d - 2.2\ Y1 \times \cos d + 3.7\ Y1 \times \sin d \qquad (8.10)$$
$$- 10.1\ Y2 \times \cos d - 12.3\ Y2 \times \sin d$$

and

$$\hat{r} = 50.9 - 13.1 \cos d - 24.0 \sin d + 6.1\ Y1 \times \cos d$$
$$- 10.3\ Y2 \times \cos d. \qquad (8.11)$$

The fitted regressions ae plotted in Figures 8.5(a) and (b); the values of R^2

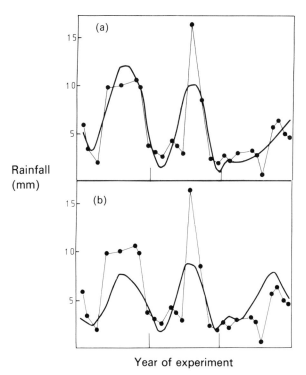

Figure 8.5   Rainfall observed during the FitzGerald experiment and that predicted by: (a) Equation (8.10); (b) Equation (8.11).

for these regressions were:

Equation (8.10): 0.67,

Equation (8.11): 0.34.

Equation (8.10) provided a good fit to the rainfall during the experiment. Therefore, adding rainfall to Equation (2.10) may not improve the fit because most of the effect of rainfall would already have been accounted for. Rainfall may, however, substitute for some of the time-variables in Equation (2.10) and thus explain the need for some of the time-variables. This was investigated in section 8.6.2 and resulted in Equation (8.6).

Equation (8.11) gave a much poorer fit to rainfall than (8.10). Therefore, rainfall would be expected to improve the fit of Equation (2.11) if it were important. This was shown to be so in section 8.7.2. Rainfall also substituted for some of the time-variables in Equation (2.11).

9 Correlation between series of random variables

9.1 INTRODUCTION

It frequently happens that more than one random variable is studied when observations are repeated on the one unit. For example, in the example of Chapter 3, Brownlee *et al.* measured available dry matter of the pasture and the nitrogen and phosphorus content of the lucerne component as well as live weight. They were interested in the effects of the treatments on each variable and also in the relations among the variables. The latter will be considered here. The method to be used is almost the same as that used for the environmental variables in Chapter 8. However, there are logical differences.

1. With random variables, each unit in the experiment can and should have an individual value for each random variable at each sampling time – an observation on an environmental variable such as rainfall applies to all units.
2. Although there may be errors in the environmental variable they will not be correlated with those in the random variable, whereas errors in the random variables are likely to be correlated.

When two variables, y_1 and y_2, are both related to a third variable, often time, the relation between y_1 and y_2 independent of time is usually required. This is given by the partial regression coefficient of y_1 on y_2 obtained by regressing y_1 on time and y_2. In some situations the partial correlation coefficient is more appropriate than the partial regression coefficient. One of the purposes of measuring dry matter in the experiment of Brownlee *et al.* (Chapter 3) was to determine the relation between live weight and dry matter. It would be appropriate to regress live weight on dry matter. However, when the relation between N and P content of the lucerne is being examined, neither variable could be considered to depend on the other and partial correlation methods would be appropriate. The methods are very similar. The partial regression coefficient, b_{ij}, or the partial correlation coefficient, r_{ij}, is calculated for each plot of the experiment which is assumed to consist of j replicates of i treatments. Then the values of b_{ij} or r_{ij} are analysed following the design of the experiment. As usual, the r_{ij} would be transformed to $z_{ij} = \tanh^{-1} r_{ij}$ before analysis (Snedecor and Cochran, 1980).

9.2 DATA

The data to be used to illustrate the method came from the experiment by Brownlee *et al.* described in Chapter 3. The live weights (L) used in Chapter 3 were recorded on 17 occasions and available dry matter (D) and the nitrogen (N) and phosphorus (P) content of the lucerne component were estimated on all but the last occasion. The data for the first 16 occasions are given in Table 9.1. Live weight in kilograms and dry matter in kg per hectare are given as $\log_e$ and N% and P% have been multiplied by 100 and 1000 respectively.

The grand means $\bar{y}_L$, $\bar{y}_D$, $\bar{y}_N$ and $\bar{y}_P$ for each occasion are plotted against time in Figure 9.1. These graphs certainly indicated strong simple correlations.

9.3 REGRESSION OF LIVE WEIGHT ON DRY MATTER

The values of y_L for each plot were regressed on the values of $\xi_1, \xi_2, \ldots, \xi_6$ for 16 equally spaced intervals, and on the values of y_D for that plot. $\xi_1, \ldots, \xi_6$ were used to remove the time trend because a sixth degree polynomial was required to fit the mean live weight in the analysis of Chapter 3. The variables for the regression for the first plot of Table 9.1 are given in Table 9.2 in the usual set-up for regression analysis.

The values of the partial regression coefficient of live weight on dry matter $b_{L(D)}$ for the 18 plots of the experiment are given in Table 9.3. The 18 values of $b_{L(D)}$ were analysed using the model of Equation (3.1), i.e.

$$\hat{b}_{L(D)} = c_0 + c_1 x_1 + c_2 x_2 + c_{22} x_2^2 + c_3 x_3 + c_{33} x_3^2 + c_{23} x_2 x_3$$
$$+ c_{233} x_2 x_3^2 + c_{222} x_2^2 x_3 + c_{2233} x_2^2 x_3^2 \tag{9.1}$$

to estimate their variance, to test the significance of their mean (c_0) and to test for effects of S and P on the partial regression coefficient.

The analysis of variance and the estimates of the mean partial regression coefficient, $\bar{b}_{L(D)}$, and of the treatment effects from the analysis of variance, are given in Table 9.4; $\bar{b}_{L(D)} = 0.0785 \pm 0.00342$. So, if a log (dry matter) of 3.06 (i.e. dry matter 21.3 kg/ha) corresponded to a log (live weight) of 3.58 (live weight 35.9 kg), an increase in log D from 3.06 to 5.48 (240 kg/ha) would be expected to increase log L by 2.42×0.0785, i.e. from 3.58 to 3.77 (live weight 43.4 kg).

The analysis of variance indicated treatment effects on $b_{L(D)}$. The significant coefficient for $x_2^2(P^2)$ indicated a quadratic relation between $b_{L(D)}$ and rate of superphosphate, i.e.

$$\hat{b}_{L(D)} = c_0 + c_2' x_2 + c_{22} x_2^2. \tag{9.2}$$

However, the significant coefficient for $x_2 x_3^2$ indicated a quadratic relation between c_2' and x_3 (section 3.3.1), i.e.

$$\hat{c}_2' = k_0 + k_1 x_3 + k_2 x_3^2. \tag{9.3}$$

Table 9.1 Values of log L, log D, $N \times 10^2$ and $P \times 10^3$ for 16 periods for the experiment of Brownlee *et al.*

Plot BPS	Period 1	2	3	4	5	6	7	8	9	10	11	12	13	14	15	16
111	3.86	3.89	3.78	3.62	3.67	3.83	3.76	3.90	3.85	3.91	3.83	3.72	3.67	3.61	3.61	3.58
	5.48	4.90	4.24	3.97	3.30	4.97	4.36	5.27	5.46	5.34	4.77	4.26	3.87	3.71	3.58	3.06
	163	176	127	465	448	316	433	219	296	388	154	112	252	133	225	119
	100	80	40	260	260	120	240	120	140	95	63	68	127	53	184	55
112	3.85	3.84	3.73	3.58	3.67	3.84	3.83	3.91	3.87	3.93	3.89	3.76	3.74	3.70	3.66	3.60
	5.75	5.36	4.57	4.20	3.55	5.31	4.94	5.33	5.80	5.64	5.03	4.53	4.32	4.09	4.15	2.91
	188	171	184	465	481	318	430	258	249	332	199	277	325	322	367	206
	130	90	70	260	290	140	250	130	140	115	97	119	179	206	258	178
113	3.91	3.92	3.81	3.66	3.79	3.89	3.89	3.98	3.99	3.98	3.93	3.78	3.81	3.71	3.70	3.55
	5.70	5.02	4.80	3.88	3.94	5.08	4.63	5.44	5.81	5.29	5.09	4.37	4.26	3.69	3.72	2.56
	188	163	134	483	502	303	470	246	288	311	143	143	318	129	343	140
	150	100	60	300	310	150	270	140	150	186	85	99	184	84	243	99
121	3.82	3.80	3.67	3.49	3.57	3.70	3.65	3.79	3.81	3.81	3.71	3.57	3.59	3.57	3.57	3.51
	5.38	4.87	4.28	3.32	3.39	4.80	4.15	5.08	5.55	5.07	4.72	3.76	3.03	2.55	2.72	2.55
	189	200	128	457	531	339	512	252	252	265	115	262	367	406	364	255
	110	70	40	240	310	120	280	110	120	102	18	166	184	312	220	169
122	3.88	3.87	3.72	3.54	3.61	3.75	3.72	3.87	3.87	3.92	3.84	3.64	3.68	3.63	3.63	3.57
	5.48	5.07	3.64	3.33	3.29	5.05	4.33	5.34	5.52	5.19	4.61	3.26	3.21	3.69	2.93	2.98
	163	222	127	469	510	323	505	278	293	336	115	185	374	308	322	304
	120	110	40	250	340	140	320	120	150	201	53	169	265	178	235	240
123	3.85	3.85	3.71	3.54	3.61	3.76	3.72	3.86	3.88	4.04	3.80	3.61	3.66	3.63	3.67	3.59
	5.45	5.26	4.38	3.32	3.71	5.23	4.81	5.63	5.72	5.30	4.90	3.88	4.34	3.48	3.59	2.95
	205	174	209	514	514	353	481	293	270	208	140	269	332	304	332	283
	130	80	80	310	300	150	310	140	140	254	62	213	255	240	235	228

Table 9.1 cont.

Plot BPS	Period															
	1															16
131	3.78	3.76	3.56	3.50	3.49	3.65	3.59	3.73	3.74	3.74	3.68	3.54	3.56	3.53	3.53	3.50
	5.64	4.82	4.30	2.77	3.07	4.74	4.49	5.12	5.42	5.15	5.02	3.64	3.83	3.47	3.84	3.20
	246	177	261	429	515	337	465	256	287	276	189	521	374	399	339	304
	120	80	80	250	280	120	250	100	110	80	55	337	235	260	203	132
132	3.84	3.78	3.62	3.50	3.47	3.67	3.64	3.78	3.80	3.79	3.71	3.55	3.60	3.58	3.59	3.55
	5.63	5.08	4.50	3.14	3.31	4.98	4.80	5.14	5.67	5.42	4.77	3.81	3.90	3.35	3.98	3.00
	164	208	134	508	496	300	462	264	278	220	112	528	332	360	343	238
	130	100	50	280	360	110	320	130	140	119	72	391	257	283	250	119
133	3.82	3.77	3.63	3.50	3.53	3.66	3.63	3.75	3.80	3.77	3.73	3.56	3.61	3.57	3.60	3.51
	5.66	5.35	4.63	3.02	3.37	4.99	4.78	5.09	5.88	5.15	5.00	3.11	4.21	3.31	4.19	2.44
	154	203	156	548	514	328	475	248	289	322	129	388	381	416	364	262
	130	100	70	330	340	160	300	120	160	156	33	312	250	305	255	120
211	3.82	3.79	3.72	3.54	3.57	3.73	3.71	3.83	3.87	3.87	3.86	3.68	3.68	3.61	3.63	3.56
	5.60	5.32	4.39	3.85	3.50	4.89	4.55	5.30	5.53	5.12	4.96	4.39	3.86	3.83	3.85	2.99
	166	191	149	484	488	317	487	274	305	269	136	136	339	136	332	259
	80	80	50	240	230	120	230	120	110	119	53	50	152	48	199	156
212	3.88	3.88	3.80	3.67	3.72	3.86	3.84	3.90	3.92	3.91	3.86	3.74	3.73	3.70	3.68	3.57
	5.73	5.55	4.53	4.13	3.66	5.37	4.59	5.40	5.98	5.61	5.23	4.46	4.62	3.55	4.13	2.87
	175	191	139	495	506	349	495	278	269	220	112	206	339	227	353	266
	100	80	40	280	280	120	270	130	120	119	78	119	191	152	211	122
213	3.91	3.91	3.84	3.72	3.82	3.88	3.88	3.91	3.90	3.93	3.91	3.77	3.81	3.74	3.74	3.66
	5.48	5.31	4.89	4.28	4.11	5.42	4.79	5.67	5.95	5.58	5.40	4.61	4.44	3.50	3.49	2.80
	209	227	171	514	488	282	501	237	298	185	178	157	346	168	381	196
	120	90	80	350	280	140	300	130	140	270	100	100	204	119	253	132

221	3.83	3.81	3.68	3.51	3.56	3.70	3.66	3.78	3.77	3.86	3.80	3.61	3.62	3.61	3.61	3.57
	5.55	4.70	4.45	3.61	3.30	5.06	4.74	5.15	5.59	5.15	4.97	4.05	3.67	3.67	3.79	3.26
	247	212	140	476	488	284	499	264	322	231	157	171	332	248	385	206
	100	70	40	230	250	110	240	100	110	68	52	102	161	152	142	137
222	3.87	3.84	3.75	3.56	3.63	3.77	3.76	3.88	3.88	3.89	3.84	3.65	3.69	3.65	3.67	3.62
	5.76	5.16	4.64	3.71	3.69	5.01	4.62	5.35	5.96	5.30	5.35	4.24	4.08	3.96	3.62	3.16
	153	170	106	513	502	267	520	223	295	171	161	185	315	133	399	255
	80	80	40	270	310	140	280	110	120	82	25	120	186	72	191	169
223	3.82	3.85	3.69	3.51	3.60	3.80	3.70	3.89	3.89	3.92	3.80	3.66	3.64	3.65	3.61	3.60
	5.51	5.18	4.44	3.92	3.41	5.14	4.62	5.52	5.68	5.28	4.89	4.39	4.05	3.73	3.19	3.44
	242	198	125	514	505	306	535	324	291	220	135	308	268	409	385	311
	150	90	60	270	310	130	310	140	140	156	50	268	245	297	213	181
231	3.76	3.78	3.56	3.45	3.47	3.62	3.51	3.72	3.74	3.76	3.59	3.48	3.52	3.52	3.51	3.52
	5.53	5.21	4.14	3.24	3.20	4.87	4.51	5.18	5.44	5.23	4.78	3.38	3.68	2.44	3.45	2.56
	193	200	153	514	519	286	512	268	314	280	224	346	374	416	336	252
	90	80	50	290	260	120	220	110	110	143	35	265	193	270	206	144
232	3.79	3.80	3.64	3.50	3.57	3.75	3.65	3.82	3.82	3.88	3.74	3.59	3.66	3.64	3.63	3.63
	5.46	5.47	3.99	3.47	3.58	5.16	4.79	5.76	5.73	5.77	4.67	4.62	3.75	3.24	3.51	3.28
	166	211	109	476	526	303	517	258	304	234	143	430	206	315	381	325
	70	90	50	270	280	110	280	110	140	101	36	324	117	275	198	220
233	3.84	3.83	3.69	3.53	3.58	3.72	3.66	3.82	3.85	3.87	3.74	3.62	3.64	3.62	3.61	3.58
	5.65	5.21	4.10	3.07	3.42	5.16	4.75	5.37	5.39	5.52	4.86	4.29	3.93	4.29	3.90	3.64
	199	194	109	515	503	279	455	274	285	240	129	381	371	381	395	220
	120	100	50	320	320	130	280	120	140	180	53	315	270	295	235	110

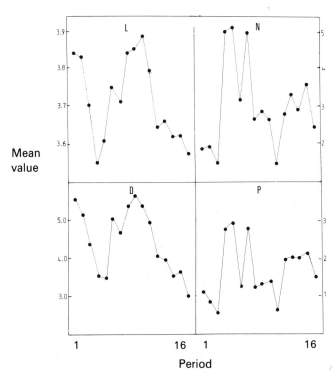

Figure 9.1 Mean values for log L, log D, N and P for 16 times of observation for Brownlee *et al.*

Table 9.2 Variables for the regression of live weight on time and dry matter for plot 1 of Table 9.1

y Live weight	ξ_1	ξ_2	ξ_3	ξ_4	ξ_5	ξ_6	Dry matter
3.86	−15	35	−455	273	−143	65	5.48
3.89	−13	21	−91	−91	143	−117	4.90
3.78	−11	9	143	−221	143	−39	4.24
3.62	−9	−1	267	−201	33	59	3.97
3.67	−7	−9	301	−101	−77	87	3.30
3.83	−5	−15	265	23	−131	45	4.97
3.76	−3	−19	179	129	−115	−25	4.36
3.90	−1	−21	63	189	−45	−75	5.27
3.85	1	−21	−63	189	45	−75	5.46
3.91	3	−19	−179	129	115	−25	5.34
3.83	5	−15	−265	23	131	45	4.77
3.72	7	−9	−301	−101	77	87	4.26
3.67	9	−1	−267	−201	−33	59	3.87
3.61	11	9	−143	−221	−143	−39	3.71
3.61	13	21	91	−91	−143	−117	3.58
3.58	15	35	455	273	143	65	3.06

Table 9.3 Values of the partial regression coefficient of live weight on dry matter for each plot of Table 9.1

B	S	1	P 2 $b_{L(D)} \times 10^3$	3
1	1	78.69	54.72	80.18
1	2	96.69	76.06	92.63
1	3	60.49	77.85	62.23
2	1	100.36	47.68	52.96
2	2	74.26	97.06	10.38
2	3	63.03	96.40	90.77

Table 9.4 Analysis of variance of the values of $b_{L(D)}$ given in Table 9.3

Source	DF	MS	F	P(%)
Blocks	1	158.2		
x_2	1	109.1	0.52	49.2
x_2^2	1	1454.2	6.9	3.0
x_3	1	20.4	0.10	76.4
x_3^2	1	110.7	0.53	48.9
$x_2 x_3$	1	710.5	3.4	10.4
$x_2 x_3^2$	1	1340.6	6.4	3.6
$x_2^2 x_3$	1	270.3	1.3	29.0
$x_2^2 x_3^2$	1	5.6	0.03	87.5
Error	8	210.5		

Variable	Estimates Coefficient	Standard error
Constant	78.47	3.42
x^2	3.02	4.19
x_2^2	−6.36	2.42
x_3	1.30	4.19
x_3^2	1.75	2.42
$x_2 x_3$	9.42	5.13
$x_2 x_3^2$	−7.47	2.96
$x_2^2 x_3$	−3.36	2.96
$x_2^2 x_3^2$	−0.28	1.71

Substituting Equation (9.3) in Equation (9.2) gave

$$\hat{b}_{L(D)} = c_0 + c_2 x_2 + c_{22} x_2^2 + c_{23} x_2 x_3 + c_{233} x_2 x_3^2.$$

Although this was the equation implied by the analysis of variance, the main effects of x_2 would usually be included as well because its interactions were needed. Thus the following equation was used to calculate the predicted values for $b_{L(D)}$:

$$\hat{b}_{L(D)} = 78.47 + 3.02x_2 - 6.36x_2^2 + 1.30x_3 + 1.75x_3^2 + 9.42x_2x_3$$
$$- 7.47x_2x_3^2. \qquad (9.4)$$

The values of $\hat{b}_{L(D)}$ for the nine combinations of S and P used in the experiment were calculated from Equation (9.4) and the values of x_2, x_2^2, x_3, x_3^2, x_2x_3 and $x_2x_3^2$ in Table 3.4 as

$$\hat{b}_{L(D)} = \begin{bmatrix} x_0 & x_2 & x_2^2 & x_3 & x_3^2 & x_2x_3 & x_2x_3^2 \\ 1 & -1 & 1 & -1 & 1 & 1 & -1 \\ 1 & -1 & 1 & 0 & -2 & 0 & 2 \\ 1 & -1 & 1 & 1 & 1 & -1 & -1 \\ 1 & 0 & -2 & -1 & 1 & 0 & 0 \\ 1 & 0 & -2 & 0 & -2 & 0 & 0 \\ 1 & 0 & -2 & 1 & 1 & 0 & 0 \\ 1 & 1 & 1 & -1 & 1 & -1 & 1 \\ 1 & 1 & 1 & 0 & -2 & 2 & -2 \\ 1 & 1 & 1 & 1 & 1 & 1 & 1 \end{bmatrix} \begin{bmatrix} 78.47 \\ 3.02 \\ -6.36 \\ 1.30 \\ 1.75 \\ 9.42 \\ -7.47 \end{bmatrix} = \begin{bmatrix} 86.43 \\ 50.65 \\ 70.20 \\ 91.64 \\ 87.69 \\ 94.24 \\ 58.69 \\ 86.57 \\ 79.13 \end{bmatrix}$$

These values are difficult to interpret in this example because the flocks were hand-fed. The flocks at $x_2 = 0$ and 1 were hand-fed as required in the winter and spring of 1971 and continually for the second half of 1972. The flocks at $x_2 = -1$ were only fed towards the end of the experiment.

9.4 CORRELATION BETWEEN NITROGEN AND PHOSPHORUS CONTENT OF LUCERNE

Both N and P displayed high-order variation with time (Figure 9.1) and again a sixth degree polynomial in time was fitted to eliminate the time trends. The variables for the bivariate regression of N and P on $\xi_1, \ldots, \xi_6$ for plot 1 are given in Table 9.5 as an example. The error sums of squares and products matrix for this analysis was

$$\begin{bmatrix} 100\,530 & 50\,507 \\ 50\,507 & 38\,928 \end{bmatrix},$$

so that

$$r_{NP} = 50\,507/(100\,530 \times 38\,928)^{1/2}$$
$$= 0.8074.$$

The values of r_{NP} for each plot of the experiment are given in Table 9.6 together with the values of $\tanh^{-1} r_{NP}$.

Table 9.5   Variables for the bivariate regression of N and P on time for data of plot 1 of Table 9.1

Y		X					
$y_1(N)$	$y_2(P)$	ξ_1	ξ_2	ξ_3	ξ_4	ξ_5	ξ_6
163	100	-15	35	-455	273	-143	65
176	80	-13	21	-91	-91	143	-117
127	40	-11	9	143	-221	143	-39
465	260	-9	-1	267	-201	33	59
448	260	-7	-9	301	-101	-77	87
316	120	5	-15	265	23	-131	45
433	240	-3	-19	179	129	-115	-25
219	120	-1	-21	63	189	-45	-75
296	140	1	-21	-63	189	45	-75
388	95	3	-19	-179	129	115	-25
154	63	5	-15	-265	23	131	45
112	68	7	-9	-301	-101	77	87
252	127	9	-1	-267	-201	-33	59
133	53	11	9	-143	-221	-143	-39
225	184	13	21	91	-91	-143	-117
119	55	15	35	455	273	143	65

The values of $z = \tanh^{-1} r_{NP}$ were analysed using the model of Equation (3.1) to estimate the variance, to test the significance of the mean and to test for effects of S and P on the partial correlation coefficient. The analysis is given in Table 9.7. There were no significant effects of treatments; in fact, all the F-values were lower than expected. The mean was

$$\bar{z} = 1.687 \pm 0.115$$

which was highly significant and corresponds to a partial correlation coefficient of 0.9338.

Table 9.6   Values of the partial correlation coefficient, r_{NP}, between N and P for each plot of Table 9.1 and $\tanh^{-1} r_{NP}$

		r_{NP}			$\tanh^{-1} r_{NP}$		
			P			P	
B	S	1	2	3	1	2	3
1	1	0.8074	0.8966	0.9847	1.120	1.455	2.433
1	2	0.9064	0.9474	0.8599	1.507	1.806	1.293
1	3	0.9635	0.9516	0.9578	1.993	1.848	1.919
2	1	0.9780	0.9493	0.7992	2.249	1.825	1.096
2	2	0.9116	0.9439	0.9226	1.537	1.773	1.606
2	3	0.8562	0.9310	09623	1.279	1.666	1.976

Table 9.7   Analysis of variance of the values of $\tanh^{-1} r_{NP}$ given in Table 9.6

Source	DF	$MS \times 10^4$	F	P(%)
Blocks	1	74.0		
x_2	1	211.3	<1	77.4
x_2^2	1	914.4	–	55.4
x_3	1	340.3	–	71.6
x_3^2	1	151.0	–	80.8
$x_2 x_3$	1	267.6	–	74.7
$x_2 x_3^2$	1	16.9	–	93.5
$x_2^2 x_3$	1	479.8	–	66.6
$x_2^2 x_3^2$	1	1171.7	–	50.4
Error	8	2394.6		

9.5 DISCUSSION

The nature of the partial regression and partial correlation established in sections 9.3 and 9.4 can be investigated in a similar manner to that used in section 8.8 for environmental variables. The values of log L, log D, N and P for each plot are regressed on ξ_1, ξ_2, ξ_3, ξ_4, ξ_5 and ξ_6 and the observed

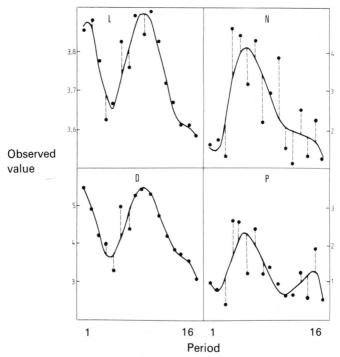

Figure 9.2   Observed values of log L, log D, N and P for plot 1 of Brownlee *et al.* on 16 occasions and values predicted by sixth degree polynomials.

and predicted values are plotted as a graph. The values for plot 1 are shown in Figure 9.2. The partial regression of log L on log D depends mainly on the variation in log D after the time-trend is removed. For plot 1 this variation is restricted to observations 4, 5, 6 and 7. The partial correlation between N and P depends on the joint variation of N and P after the time-trend is removed. Figure 9.2 indicates that this variation covers the whole period for plot 1.

The quantity z used in section 9.4 is very important in correlation analysis; see, for example, sections 10.6 and 17.16 of Snedecor and Cochran (1980). When z is calculated for n pairs of observations taken on variables which follow a bivariate normal distribution, z is close to normal and has variance $(n-3-p)^{-1}$; p is the number of variables eliminated when partial correlation is used. In the example in this chapter, $n = 16$ and $p = 6$, so that

$$\text{var}(z) = 1/7$$
$$= 0.1429.$$

Var(z) was estimated as 0.23946 in the analysis of variance of Table 9.7. The difference between the two estimates probably indicates that the correlation differs somewhat from plot to plot, i.e. as usual in field experiments there is a component of variance for plots. The observed variance, 0.23946, is the appropriate one to use for tests of significance and to calculate standard errors.

10 Response (reaction) times

10.1 INTRODUCTION

Individual units were measured repeatedly in the experiments considered in the previous chapters. A completely different type of experiment will be considered in this chapter. Characteristically, the response of an individual can only be measured once. For example, the response may be the time taken for a fruit to ripen after harvest or the time taken for an insect to die after being exposed to an insecticide. Each unit in the experiment consists of a sample of individuals and the response is observed for each individual. The unit is usually assessed repeatedly so that the data are produced serially.

Experiments of this type are very common in toxicology. Although they are not as common as field or pen trials in agricultural and general biological research, they occur frequently. Examples include studies on seed germination, vase-life of cut flowers and shelf-life of fruit and vegetables as well as toxicity in entomology and plant pathology. Studies on storage life of fruit in cool storage produce a different type of data because different samples must be removed from storage and ripended after each storage period; regression analysis based on the probit or logit transformation are applicable for the quantal data usually obtained but this will not be considered here.

A two-stage analysis is again convenient for response times:

1. Stage 1 (within units) – estimate the mean response and possibly the variance.
2. Stage 2 (between units) – analyse the values of the mean and variance obtained in Stage 1 in the design of the experiment to estimate the treatment effects and to test their significance.

Two aspects of experimental design correspond to the two stages in the analysis. The selection of the units and the application of the treatments to them are based on the usual principles of experimental design. This must include replication of the treatments to provide the appropriate error term to test treatment effects. The procedures used to measure the individuals in a unit, i.e. the selection of intervals, and to analyse these measurements are devised to estimate their mean and variance efficiently without bias. The methods usually used in the within-unit analysis are based on estimation in

Example   207

the normal distribution. The application of these methods to response times in general was considered in detail by Sampford (1952a, 1952b, 1954); their application to seed germination tests was considered more recently by Hunter *et al.* (1984).

10.2 EXAMPLE

An example of the type of data is given in Table 10.1. Most of the important features can be illustrated with this small set of data. Six batches of insects were sprayed with insecticide, three sets with compound A and three with compound B. The number of insects knocked down was counted every 2 minutes up to 14 minutes. This procedure automatically grouped the response times into 2-minute classes. Grouping is usually unavoidable in this type of work.

The data for the individual batches for each insecticide were combined initially to study the distribution of the knock-down times because each batch was small and the sets of counts of each insecticide did not appear much different. This may not always be possible and, in fact, counts from different sets should not be combined unless there are obviously no differences.

10.2.1 Distribution of knock-down times

The frequencies in the intervals for each insecticide have been used to construct the bar diagrams in Figure 10.1. These diagrams are like histograms of a normal distribution, but appear to be skewed. A more sensitive guide is to plot the data in a form that can be fitted by a straight line if the distribution is normal. The procedure, which can be seen in Table 10.1, follows.

1. Form the cumulative frequencies.
2. Convert the cumulative frequencies to percentages of the total number of insects.
3. Convert the percentages to normal equivalent deviates (NEDs) from tables of the cumulative normal distribution.
4. Plot the deviates against the upper limit of the classes. This procedure is available in most computer statistical programs.

If the distribution is normal, a straight line will result. If not, it may be possible to transform the abscissa (t) to obtain a straight line; log t is by far the most common transformation, followed by $1/t$.

The variable that is normally distributed is called a metameter. The NEDs for insecticides A and B are plotted against t and against log t in Figure 10.2. The points lie close to the straight lines in Figure 10.2(b), and log t will be

Table 10.1 Number of insects knocked down in two-minute intervals for six batches and the results of some calculations on these

Batch	Compound	Number tested	Number knocked-down Interval (minutes)						
			0–2	2.1–4	4.1–6	6.1–8	8.1–10	10.1–12	12.1–14
1	A	46	0	15	18	8	4	1	0
2	A	40	0	6	7	9	5	2	1
3	A	32	1	4	5	4	6	3	6
Sum		108	1	25	30	21	15	6	7
Cumulative sum%			1	26	56	77	92	98	105
			0.93	24.07	51.85	71.30	85.18	90.74	97.22
Normal equivalent deviate			−2.3656	−0.7041	0.0464	0.5622	1.0441	1.3249	1.9110
4	B	41	7	17	13	0	0	0	1
5	B	48	0	17	15	3	6	4	2
6	B	33	4	13	9	6	0	0	0
Sum		122	11	47	37	9	6	4	3
Cumulative sum%			11	58	95	104	110	114	117
			9.02	47.54	77.87	85.24	90.16	93.44	95.90
Normal equivalent deviate			−1.3396	−0.0617	0.7678	1.0468	1.2906	1.5094	1.7392

Example   209

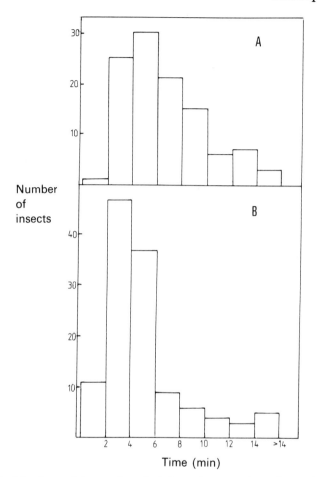

Figure 10.1 Numbers of insects knocked-down in two-minute intervals for two insecticides, A and B.

assumed to be normally distributed. It is important to note that the plots are only used to indicate which metameter should be used. Regression is not used in the formal analysis. This has been stated many times but the probit regression line is still often fitted formally although the method is completely inappropriate. The probability model for probit analysis assumes independence between successive observations.

The fact that lines in Figure 10.2(b) do not appear to be parallel indicates that the variances of the distributions may differ. This will be checked later.

This preliminary analysis indicated that the distribution of knock-down times could be approximated by a normal distribution in $\log t = x$. Thus, the probability $p(x)$ that an insect is knocked down in an interval of length dx

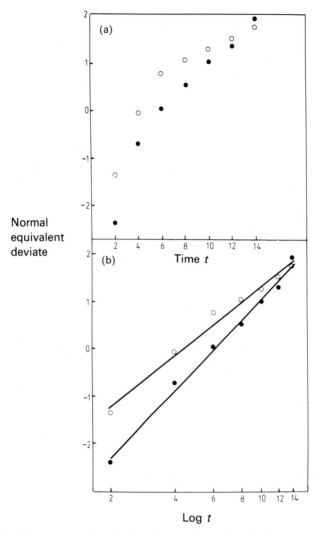

Figure 10.2   Normal equivalent deviate for cumulative knock-down against (a) time, (b) log (time) for two insecticides.

is given by

$$p(x) = \frac{1}{\sqrt{(2\pi\sigma)}} \exp[-(x-\mu)^2/2\sigma^2] \, \mathrm{d}x$$

where μ is the mean and σ^2 is the variance between insects.

Example 211

10.2.2 Initial estimates of μ and σ^2

The initial estimates, $\bar{x}_0$ and s^2, of μ and σ^2 are obtained by elementary means in which all the observations, n_i, in a class are given the value of the mid-point, x_i, of the class. Then

$$x_0 = \sum n_i x_i \Big/ \sum n_i,$$

and

$$s_0^2 = \left\{ \sum n_i x_i^2 - \left(\sum n_i x_i \right)^2 \Big/ \sum n_i \right\} \Big/ (n-1)$$

The procedure is set out in Table 10.2. Following Sampford (1952a) the mid-points were taken where possible as $\log (t_{i-1} + t_i)/2$ rather than $\frac{1}{2}(\log t_{i-1} + \log t_i)$. The mid-point for the first class is $\log t_1$ and that for the last is the second-last mid-point plus the width of the second-last class. The estimate of s_0^2 is adjusted by Sheppard's adjustment for grouping as

$$s_0^2(\text{adj}) = s_0^2 - \left(\sum n_i i^2 \right) \Big/ 12 \sum n_i,$$

where i is the width of the ith class. The values of $\bar{x}_0$ and $s_0^2(\text{adj})$ for each batch are given in Table 10.3. These are used in the next section to obtain the final estimates. The final estimates are also given in Table 10.3.

The initial estimates for batches 1 and 2 are very close to the final ones. Knock-down times were observed for each individual in these batches, whereas some knock-down times fell in the tails of the distribution for the other batches. A knock-down time cannot be assigned to observations in the tails although limits can be placed on them, i.e. they are less than the lowest class limit or greater than the highest.

Tables 10.2 Class limits and values used to calculate initial estimates of the mean and variance of knock-down times for the six sets of counts of Table 10.1

Class limits		Mid-point	Width	Batch					
t	$\log t(x)$	x_i	i	1	2	3	4	5	6
0–2	0.6931	0.6931*	–	0	0	1	7	0	4
>2–4	1.3863	1.0986	0.6932	15	6	4	17	17	13
>4–6	1.7918	1.6094	0.4055	18	7	5	13	15	9
>6–8	2.0794	1.9459	0.2876	8	9	4	0	3	6
>8–10	2.3025	2.1972	0.2231	4	5	6	0	6	0
>10–12	2.4849	2.3979	0.1824	1	2	3	0	4	0
>12–14	2.6391	2.5649	0.1542	0	1	6	1	2	0
>14		2.7191*		0	0	3	3	1	1

*Working values.

Table 10.3 Initial and final estimates of the mean and variance of knock-down times for the counts of Table 10.1

Batch	Compound	$\bar{x}_0$	μ	$s_0^2(adj)$	σ^2
1	A	1.5696	1.5765	0.129973	0.134689
2	A	1.7905	1.7985	0.174723	0.166162
3	A	2.0263	2.0426	0.308101	0.371064
4	B	1.3457	1.2654	0.273938	0.488573
5	B	1.6516	1.6532	0.241705	0.266266
6	B	1.3919	1.3413	0.197412	0.299001

10.2.3 Maximum likelihood estimates of μ and σ^2

Efficient estimates of μ and σ^2 can be obtained by the method of maximum likelihood.

Denote by P_i the probability that an observation falls in class i. Then the probability that n_i observations fall in class i is $P_i^{n_i}$.

The probability of observing the sample obtained is then π, where,

$$\log \pi = \log (N!/n_1!n_2! \ldots n_m!) + \sum_{i=1}^{m} n_i \log P_i,$$

$N = \sum_{i=1}^{m} n_i$ is a constant and m is the number of classes.

Since $n_1, n_2, \ldots, n_m$ are determined by the frequencies in the classes and are fixed, $\log (N!/n_1!n_1! \ldots n_m!)$ is a constant. Therefore, we need only consider the function

$$L = \sum_{i=1}^{m} n_i \log P_i$$

where

$$P_i = \Phi(z_i) - \Phi(z_{i-1}),$$

$$z_i = (x_i - \mu)/\sigma$$

and $\Phi(z)$ is the standard normal integral from $-\infty$ to z; x_i is the upper limit of class i now not the mid-point.

Values of μ and σ^2 that maximize L can be found by calculating the value of L for a grid of values of μ and σ^2 around the initial estimates or by other numerical methods. Computer programs are available for these. Estimates of the variance of μ and of σ^2 are obtained in these programs.

The example was selected to show how the observations in the tails are taken into account. This method can be very useful in long-term experiments. Some individuals may have very large response times, particularly when response times are log normally distributed. The method of maximum

Example   213

likelihood allows preliminary analyses to be carried out before all individuals have responded.

Although maximum likelihood provides variances for μ and for σ^2 these are not the appropriate variances for testing the difference between the estimates for the two insecticides. The variances are estimates of the variation within a batch. Insecticides are compared on different batches. The variation between the estimates for the batches of insects is needed. This aspect will be considered further in section 10.2.5 where the mean knock-down times are compared. The variances will be considered first.

10.2.4 Variances of knock-down time

The variances estimated for each batch of insects are given in Table 10.3. Variances are usually transformed to logarithms before analysis to stabilize their variance because the variance of s^2 is proportional to the square of its expected value. The transformed values and the analysis of variance are given in Table 10.4. There is no indication that the variances differ and it is appropriate to compare the means.

10.2.5 Mean knock-down time

The mean knock-down times (μ) estimated for each batch of insects by maximum likelihood and the analysis of variance of these are given in Table 10.5.

The mean square for the difference between insecticides was non-significant when compared to the between-batches mean square. This was not a very sensitive test because there were only four degrees of freedom between batches. However, it is important to note that this is the appropriate error, not the estimate based on the variation between insects, nor the asymptotic

Table 10.4

| | $\log_e \sigma^2$ | |
Insecticide A		Insecticide B
-2.005		-0.716
-1.795		-1.323
-0.991		-1.207

Source	DF	SS	MS	F	P(%)
Between insecticides	1	0.396547	0.396547	2.0	0.22
Between batches	4	0.778522	0.194631		
Total	5	1.175070			

Table 10.5

| | μ | |
Insecticide A		Insecticide B
1.5765		1.2654
1.7985		1.6532
2.0426		1.3413

Source	DF	SS	MS	F	P(%)
Between insecticides	1	0.223378	0.223378	4.62	0.98
Between batches	4	0.193183	0.048296		
Total	5	0.416561			

estimate from maximum likelihood. The average mean knock-down time, $\bar{\mu}$, was 1.6129 with standard error $(0.048296/6)^{1/2} = \pm 0.0897$.

This would be reported as 1.61 ± 0.0897. The retransformed value, $e^{1.6129} = 5.02$ is an estimate of median knock-down time, t_{50}, in minutes. A standard error cannot be assigned to t_{50} but approximate confidence limits can be obtained by obtaining confidence limits for $\bar{\mu}$ and retransforming these as follows:

$$\text{Confidence limits for } \bar{\mu} = 1.6129 \pm 0.0897 \times 2.7764$$
$$= 1.3639, 1.8619,$$
$$\text{limits for } t_{50} = e^{1.3639}, e^{1.8619}$$
$$= 3.91, 6.44 \text{ minutes}$$

(2.7754 is the 5% critical value for the t-distribution on four degrees of freedom).

10.2.6 Response time for 90%

The insecticides may be compared using statistics other than the mean. The time when 90% of individuals have responded, t_{90}, is often quoted in this type of study. This value and its confidence limits can be calculated in a similar way to those for the median, t_{50}.

The normal deviate corresponding to 90% is found from tables to be 1.2816. Therefore the log (time), x_{90}, at which 90% are knocked-down is

$$x_{90} = \mu + 1.2816\sigma.$$

The value of x_{90} for each batch was calculated using the values of μ and σ^2 given in Table 10.3. The values obtained are given in Table 10.6 together

Table 10.6

| | Estimates of x_{90} | |
	Insecticide A	*Insecticide B*
	2.0468	2.1612
	2.3209	2.3145
	2.8233	2.0421
Mean	2.3970	2.1726
s^2	0.15508	0.01865

$$\text{Mean } s^2 \text{ on 4DF} = 0.086865$$
$$s^2(\text{mean } x_{90}) = 0.086865/3$$
$$= 0.028955$$
$$s \times t_{0.05,\,4} = 0.028955^{1/2} \times 2.7764$$
$$= \pm 0.4724$$

with their means and variances and the 95% confidence limits for insecticides A and B are given in Table 10.7.

The variance of the mean was estimated to be 0.04829 in section 10.2.5 and the variance of x_{90} was just estimated to be 0.086865. Therefore, it would be very inefficient to compare treatments using the values of x_{90}; x_{90} and t_{90} should be regarded purely as descriptive statistics.

10.3 DISCUSSION

10.3.1 Grouping

Before the electronic computer, observations were grouped routinely to reduce the labour in calculating summary statistics for many types of large data-sets. Thus the effect of grouping on the precision of estimates has been studied extensively. In experiments like that considered here, grouping occurs automatically and unavoidably, but the same principles apply. The basic rules are:

1. intervals should be equal;
2. width should not exceed the standard deviation, σ, of the observations; preferably it should not exceed $\sigma/2$;
3. about 18–20 classes should be formed, certainly not less than 10.

Table 10.7

| *Insecticide A* | | *Insecticide B* | |
x_{90}	t_{90}(min)	x_{90}	t_{90}(min)
1.9246	6.85	1.7002	5.48
2.8694	17.63	2.6450	14.08

In the example considered here and in most experiments of this type, the analysis is in terms of $\log t$. The rules then apply to $\log t$. Thus, intervals should be equally spaced on a log scale. However, the investigator only needs to be able to estimate the likely range of observations to calculate appropriate intervals. The design and the data in Table 10.1 suggest that the researcher expected response times in the range 1–14 minutes. This corresponds to a range 0–2.639 in $\log_e$. With so few insects in each batch, ten classes would probably be the most aimed at. This leads to the upper-class limits of Table 10.8.

Rule 2 implies knowledge of σ which will not usually be known. However, there is a well-known and tabled relation between σ and the range in normal samples which can often be used to estimate σ. The relation depends on the number (n) of observations, e.g.

n	40	50	60	70	80	90	100
σ as fraction of range	0.231	0.222	0.216	0.210	0.216	0.206	0.202

In the example, the average number of insects was 40 and the range was guessed to be 2.639, so that σ should be around $0.231 \times 2.639 = 0.610$. Therefore, Rule 2 implies that a class interval of 0.3 would be appropriate. The width of the intervals given in Table 10.8 is 0.26 and they would be accepted on statistical grounds, but may need to be changed if they are not practical. Nevertheless, intervals should be much smaller at the start than at the end of the study when response is log normally distributed.

10.3.2 Replication

Although each individual in a unit contributes an observation such as a knock-down time, the batch of individuals makes up the basic unit to which

Table 10.8

Class	Upper limit $\log_e$	Upper limit t	Upper limit min	s
1	0.264	1.30	1	18
2	0.528	1.70	1	42
3	0.792	2.21	2	13
4	1.056	2.87	2	52
5	1.320	3.75	3	45
6	1.584	4.87	4	52
7	1.848	6.35	6	21
8	2.112	8.26	8	16
9	2.376	10.74	10	44
10	2.640	14.01	14	1
11	2.893	18.05	18	3

different treatments are applied. Therefore, batches must be replicated to provide an estimate of between-batches error because this will be the appropriate error for testing treatment effects.

Sufficient replication should be used to provide about 12 degrees of freedom for error; certainly estimates of error based on fewer than 8 degrees of freedom would not be considered to provide conclusive results by most research workers.

10.3.3 Thresholds

Commonly, there is a period after the start of an experiment during which no responses can occur. For example, seeds cannot germinate until water has been absorbed. If t_0 is the time when a response is first possible, the response time is $t - t_0$ or $\log(t - t_0)$ or similar.

The three-parameter normal or log normal distribution is then the appropriate distribution for response time. Although t_0 can be estimated by maximum likelihood as well as μ and σ^2, it is usually estimated very imprecisely. Fortunately, t_0 can usually be ignored in comparative experiments because the estimates of the percentiles of the distribution such as the mean and the 90% value are largely unaffected by the choice of t_0 (Johnson and Kotz, 1970). However, if an estimate of t_0 were available from previous work, $t - t_0$ or $\log(t - t_0)$ would be the appropriate response time.

10.3.4 General discussion

Only the simplest kind of distribution for response times has been considered. A common complication occurs when a proportion of individuals fails to respond. For example, a proportion, p, of seed may not be viable. This is not a serious complication because the method of maximum likelihood allows p to be estimated as well as μ and σ^2. Methods of doing this were discussed by Sampford (1954) in general, and more recently by Hunter et al. (1984) for seed germination. A more difficult analysis is encountered when the distribution being studied is really a mixture of distributions, as would occur if more than one stimulus resulted in the same response. McLachlan and Jones (1988) considered this type of problem recently for grouped and censored data.

An alternative method of analysis is to regard the response variable as a categorical variable with the categories defined by the intervals, and to fit a generalized linear model to the design variables (McCullagh and Nelder, 1989). This approach leads to an analysis of deviance which has many of the features of an analysis of variance. There appear to be two disadvantages with this approach in experimental work. Firstly, tests of significance and the variances of estimates are based on the sampling variances, whereas there are usually other sources of variation as well in experiments. Solutions have been proposed for this problem but they involve additional assumptions and complicate the analysis. This is an active field of research referred to as over-dispersion in generalized linear models. Secondly, it appears that the method can only be applied to experiments that have one error term so that data from split-plot designs could not be analysed in this way.

These considerations prompted the detailed discussion on grouping in section 10.3.1. Grouping and censoring cause some information to be lost and make maximum likelihood or similar analyses necessary. However, the amount of information lost can be minimized by careful design and experimentation along the lines suggested. Then, advanced methods are not needed, as the results for batches 1 and 2 in section 10.2.2 show. The estimates given by maximum likelihood differed little from the initial estimates for these batches because all the insects could be assigned bounded response times.

Appendix
Orthogonal multipliers for polynomial regression through the origin for n = 2–24

Order of polynomial coefficient

n = 2

	2
1	−0.7071
2	0.7071

n = 3

	2	3
1	−0.7071	0.4082
2	0	−0.8165
3	0.7071	0.4082

n = 4

	2	3	4
1	−0.6708	0.5	−0.2236
2	−0.2236	−0.5	0.6708
3	0.2236	−0.5	−0.6708
4	0.6708	0.5	0.2236

n = 5

	2	3	4	5
1	−0.6324	0.5345	−0.3162	0.1195
2	−0.3162	−0.2673	0.6324	−0.4781
3	0	−0.5345	0	0.7171
4	0.3162	−0.2673	−0.6324	−0.4781
5	0.6324	0.5345	0.3162	0.1195

n = 6

	2	3	4	5	6
1	−0.5976	0.5455	−0.3727	0.1890	−0.0630
2	−0.3586	−0.1091	0.5217	−0.5669	0.3150
3	−0.1195	−0.4364	0.2981	0.3780	−0.6299
4	0.1195	−0.4364	−0.2981	0.3780	0.6299
5	0.3586	−0.1091	−0.5217	−0.5669	−0.3150
6	0.5976	0.5455	0.3727	0.1890	0.0630

Order of polynomial coefficient

$n = 7$

	1	2	3	4	5	6	7
1		-0.5669	0.5455	-0.4082	0.2417	-0.1091	0.0329
2		-0.3780	0	0.4082	-0.5641	0.4364	-0.1974
3		-0.1890	-0.3273	0.4082	0.0806	-0.5455	0.4935
4		0	-0.4364	0	0.4835	0	-0.6580
5		0.1890	-0.3273	-0.4082	0.0806	0.5455	0.4935
6		0.3780	0	-0.4082	-0.5641	-0.4364	-0.1974
7		0.5669	0.5445	0.4082	0.2417	0.1091	0.0329

$n = 8$

	1	2	3	4	5	6	7	8
1		-0.5401	0.5401	-0.4308	0.2820	-0.1498	0.0615	-0.0171
2		-0.3858	0.0772	0.3077	-0.5238	0.4922	-0.3077	0.1195
3		-0.2314	-0.2314	0.4308	-0.1290	-0.3638	0.5539	-0.3585
4		-0.0772	-0.3858	0.1846	0.3626	-0.3210	-0.3077	0.5974
5		0.0772	-0.3858	-0.1846	0.3626	0.3210	-0.3077	-0.5974
6		0.2314	-0.2314	-0.4308	-0.1209	0.3638	0.5539	0.3585
7		0.3858	0.0772	-0.3077	-0.5238	-0.4922	-0.3077	-0.1195
8		0.5401	0.5401	0.4308	0.2820	0.1498	0.0615	0.0171

$n = 9$

1	2	3	4	5	6	7	8	9
1	−0.5164	0.5318	−0.4449	0.3129	−0.1849	0.0899	−0.0341	0.0088
2	−0.3873	0.1330	0.2225	−0.4693	0.5085	−0.3820	0.2048	−0.0705
3	−0.2582	−0.1519	0.4132	−0.2458	−0.1849	0.4944	−0.4780	0.2468
4	−0.1291	−0.3229	0.2860	0.2011	−0.4160	0.0225	0.4780	−0.4936
5	0	−0.3799	0	0.4023	0	−0.4495	0	0.6170
6	0.1291	−0.3229	−0.2860	0.2011	0.4160	0.0225	−0.4780	−0.4936
7	0.2582	−0.1519	−0.4132	−0.2458	0.1849	0.4944	0.4780	0.2468
8	0.3873	0.1330	−0.2225	−0.4693	−0.5085	−0.3820	−0.2048	−0.0705
9	0.5164	0.5318	0.4449	0.3129	0.1849	0.0899	0.0341	0.0088

$n = 10$

1	2	3	4	5	6	7	8	9	10
1	−0.4954	0.5222	−0.4534	0.3366	−0.2148	0.1168	−0.0527	0.0187	−0.0045
2	−0.3853	0.1741	0.1511	−0.4114	0.5013	−0.4282	0.2752	−0.1309	0.0408
3	−0.2752	−0.0870	0.3778	−0.3179	−0.0358	0.3892	−0.5035	0.3740	−0.1633
4	−0.1651	−0.2611	0.3347	0.0561	−0.3939	0.2335	0.2459	−0.5236	0.3810
5	−0.0550	−0.3482	0.1296	0.3366	−0.2148	−0.3114	0.3279	0.2618	−0.5714
6	0.0550	−0.3482	−0.1296	0.3366	0.2148	−0.3114	−0.3279	0.2618	0.5714
7	0.1651	−0.2611	−0.3347	0.0561	0.3939	0.2335	−0.2459	−0.5236	−0.3810
8	0.2752	−0.0870	−0.3778	−0.3179	0.0358	0.3892	0.5035	0.3740	0.1633
9	0.3853	0.1741	−0.1511	−0.4114	−0.5013	−0.4282	−0.2752	−0.1309	−0.0408
10	0.4954	0.5222	0.4534	0.3366	0.2148	0.1168	0.0527	0.0187	0.0045

$n = 11$

Order of polynomial coefficient

	1	2	3	4	5	6	7	8	9	10
1		−0.4767	0.5121	−0.4580	0.3548	−0.2402	0.1416	−0.0717	0.0303	−0.0101
2		−0.3814	0.2048	0.0916	−0.3548	0.4804	−0.4532	0.3298	−0.1881	0.0811
3		−0.2860	−0.0341	0.3359	−0.3548	0.0801	0.2738	−0.4733	0.4429	−0.2738
4		−0.1907	−0.2048	0.3512	−0.0591	−0.3202	0.3399	0.0287	−0.4125	0.4868
5		−0.0953	−0.3072	0.2137	0.2365	−0.3202	−0.1133	0.4016	−0.0849	−0.4259
6		0	−0.3414	0	0.3548	0	−0.3776	0	0.4247	0
7		0.0953	−0.3072	−0.2137	0.2365	0.3202	−0.1133	−0.4016	−0.0849	0.4259
8		0.1907	−0.2048	−0.3511	−0.0591	0.3202	0.3399	−0.0287	−0.4125	−0.4868
9		0.2860	−0.0341	−0.3359	−0.3548	−0.0801	0.2738	0.4733	0.4429	0.2738
10		0.3814	0.2048	−0.0916	−0.3548	−0.4804	−0.4532	−0.3298	−0.1881	−0.0811
11		0.4767	0.5121	0.4580	0.3548	0.2402	0.1416	0.0717	0.0303	0.0101

$n = 12$

	1	2	3	4	5	6	7	8	9	10
1		−0.4599	0.5018	−0.4599	0.3688	−0.2616	0.1642	−0.0905	0.0431	−0.0172
2		−0.3763	0.2281	0.0418	−0.3017	0.4519	−0.4627	0.3701	−0.2389	0.1236
3		−0.2927	0.0091	0.2927	−0.3688	0.1665	0.1642	−0.4129	0.4660	−0.3552
4		−0.2091	−0.1551	0.3484	−0.1453	−0.2299	0.3732	−0.1365	−0.2545	0.4741
5		−0.1254	−0.2646	0.2648	0.1341	−0.3488	0.0597	0.3356	−0.2898	−0.1596
6		−0.0418	−0.3193	0.0976	0.3129	−0.1586	−0.2985	0.2303	0.2741	−0.3286
7		0.0418	−0.3193	−0.0976	0.3129	0.1586	−0.2985	−0.2303	0.2741	0.3286
8		0.1254	−0.2646	−0.2648	0.1341	0.3488	0.0597	−0.3356	−0.2898	0.1596
9		0.2091	−0.1551	−0.3484	−0.1453	0.2299	0.3732	0.1365	−0.2545	−0.4741
10		0.2927	0.0091	−0.2927	−0.3688	−0.1665	0.1642	0.4129	0.4660	0.3552
11		0.3763	0.2281	−0.0418	−0.3017	−0.4519	−0.4627	−0.3701	−0.2389	−0.1236
12		0.4599	0.5018	0.4599	0.3688	0.2616	0.1642	0.0905	0.0431	0.0172

$n = 13$

1	2	3	4	5	6	7	8	9	10
1	−0.4447	0.4917	−0.4599	0.3794	−0.2797	0.1845	−0.1086	0.0564	−0.0254
2	−0.3706	0.2458	0	−0.2530	0.4195	−0.4614	0.3981	−0.2820	0.1653
3	−0.2965	0.0447	0.2509	−0.3680	0.2288	0.0671	−0.3389	0.4563	−0.4068
4	−0.2224	−0.1117	0.3345	−0.2070	−0.1398	0.3607	−0.2468	−0.0974	0.3941
5	−0.1482	−0.2235	0.2927	0.0422	−0.3305	0.1845	0.2138	−0.3640	0.0763
6	−0.0741	−0.2905	0.1672	0.2453	−0.2542	−0.1678	0.3290	0.0513	−0.3814
7	0	−0.3129	0	0.3220	0	−0.3355	0	0.3589	0
8	0.0741	−0.2905	−0.1672	0.2453	0.2542	−0.1678	−0.3290	0.0513	0.3814
9	0.1482	−0.2235	−0.2927	0.0422	0.3305	0.1845	−0.2138	−0.3640	−0.0763
10	0.2224	−0.1117	−0.3345	−0.2070	0.1398	0.3607	0.2468	−0.0974	−0.3941
11	0.2965	0.0447	−0.2509	−0.3680	−0.2288	0.0671	0.3389	0.4563	0.4068
12	0.3706	0.2458	0	−0.2530	−0.4195	−0.4614	−0.3981	−0.2820	−0.1653
13	0.4447	0.4917	0.4599	0.3794	0.2797	0.1845	0.1086	0.0564	0.0254

$n = 14$

1	2	3	4	5	6	7	8	9	10
1	−0.4309	0.4818	−0.4586	0.3876	−0.2949	0.2028	−0.1257	0.0699	−0.0344
2	−0.3646	0.2594	−0.0353	−0.2087	0.3856	−0.4523	0.4159	−0.3173	0.2041
3	−0.2983	0.0741	0.2116	−0.3578	0.2722	−0.0156	−0.2612	0.4248	−0.4321
4	−0.2320	−0.0741	0.3143	−0.2493	−0.0577	0.3218	−0.3104	0.0376	0.2836
5	−0.1657	−0.1853	0.3046	−0.0352	−0.2866	0.2623	0.0835	−0.3495	0.2359
6	−0.0994	−0.2594	0.2148	0.1707	−0.2990	−0.0354	0.3297	−0.1344	−0.2783
7	−0.0331	−0.2965	0.0770	0.2927	−0.1237	−0.2836	0.1759	0.2689	−0.2386
8	0.0331	−0.2965	−0.0770	0.2927	0.1237	−0.2836	−0.1759	0.2689	0.2386
9	0.0994	−0.2594	−0.2148	0.1707	0.2990	−0.0354	−0.3297	−0.1344	0.2783
10	0.1657	−0.1853	−0.3046	−0.0352	0.2866	0.2623	−0.0835	−0.3495	−0.2359
11	0.2320	−0.0741	−0.3143	−0.2493	0.0577	0.3218	0.3104	0.0376	−0.2836
12	0.2983	0.0741	−0.2116	−0.3578	−0.2722	−0.0156	0.2612	0.4248	0.4321
13	0.3646	0.2594	0.0353	−0.2087	−0.3856	−0.4523	−0.4159	−0.3173	−0.2041
14	0.4309	0.4818	0.4586	0.3876	0.2949	0.2028	0.1257	0.0699	0.0344

$$n = 15$$

Order of polynomial coefficient

	1	2	3	4	5	6	7	8	9	10
1		−0.4183	0.4723	−0.4562	0.3936	−0.3077	0.2190	−0.1418	0.0833	−0.0440
2		−0.3586	0.2699	−0.0652	−0.1687	0.3517	−0.4380	0.4256	−0.3452	0.2391
3		−0.2988	0.0986	0.1755	−0.3417	0.3010	−0.0842	−0.1855	0.3799	−0.4360
4		−0.2390	−0.0415	0.2908	−0.2768	0.0135	0.2695	−0.3383	0.1437	0.1665
5		−0.1793	−0.1505	0.3058	−0.0979	−0.2309	0.3017	−0.0327	−0.2847	0.3189
6		−0.1195	−0.2284	0.2457	0.0987	−0.3074	0.0766	0.2728	−0.2518	−0.1210
7		−0.0598	−0.2750	0.1354	0.2442	−0.2075	−0.1914	0.2728	0.1144	−0.3267
8		0	−0.2906	0	0.2973	0	−0.3063	0	0.3204	0
9		0.0598	−0.2750	−0.1354	0.2442	0.2075	−0.1914	−0.2728	0.1144	0.3267
10		0.1195	−0.2284	−0.2457	0.0987	0.3074	0.0766	−0.2728	−0.2518	0.1210
11		0.1793	−0.1505	−0.3058	−0.0979	0.2309	0.3017	0.0327	−0.2847	−0.3189
12		0.2390	−0.0415	−0.2908	−0.2768	−0.0135	0.2695	0.3383	0.1437	−0.1665
13		0.2988	0.0986	−0.1755	−0.3417	−0.3010	−0.0842	0.1855	0.3799	0.4360
14		0.3586	0.2699	0.0652	−0.1687	−0.3517	−0.4380	−0.4256	−0.3452	−0.2391
15		0.4183	0.4723	0.4562	0.3936	0.3077	0.2190	0.1418	0.0833	0.0440

$$n = 16$$

1	2	3	4	5	6	7	8	9	10
1	−0.4067	0.4631	−0.4532	0.3981	−0.3185	0.2334	−0.1569	0.0964	−0.0539
2	−0.3525	0.2778	−0.0906	−0.1327	0.3185	−0.4202	0.4288	−0.3664	0.2697
3	−0.2983	0.1191	0.1424	−0.3223	0.3185	−0.1401	−0.1150	0.3278	−0.4238
4	−0.2440	−0.0132	0.2660	−0.2931	0.0735	0.2119	−0.3403	0.2210	0.0563
5	−0.1898	−0.1191	0.2998	−0.1473	−0.1715	0.3125	−0.1263	−0.1973	0.3408
6	−0.1356	−0.1985	0.2640	0.0335	−0.2918	0.1616	0.1890	−0.3041	0.0314
7	−0.0813	−0.2514	0.1783	0.1881	−0.2562	−0.0898	0.3017	−0.0371	−0.2993
8	−0.0271	−0.2778	0.0628	0.2756	−0.1002	−0.2694	0.1408	0.2596	−0.1867
9	0.0271	−0.2778	−0.0628	0.2756	0.1002	−0.2694	−0.1408	0.2596	0.1867
10	0.0813	−0.2514	−0.1783	0.1881	0.2562	−0.0898	−0.3017	−0.0371	0.2993
11	0.1356	−0.1985	−0.2640	0.0335	0.2918	0.1616	−0.1890	−0.3041	−0.0314
12	0.1898	−0.1191	−0.2998	−0.1473	0.1715	0.3125	0.1263	−0.1973	−0.3408
13	0.2440	−0.0132	−0.2660	−0.2931	−0.0735	0.2119	0.3403	0.2210	−0.0563
14	0.2983	0.1191	−0.1424	−0.3223	−0.3185	−0.1401	0.1150	0.3278	0.4238
15	0.3525	0.2778	0.0906	−0.1327	−0.3185	−0.4202	−0.4288	−0.3664	−0.2697
16	0.4067	0.4631	0.4532	0.3981	0.3185	0.2334	0.1569	0.0964	0.0539

Order of polynomial coefficient

$n = 17$

1	2	3	4	5	6	7	8	9	10
1	−0.3960	0.4543	−0.4497	0.4012	−0.3276	0.2463	−0.1708	0.1091	−0.0640
2	−0.3466	0.2839	−0.1124	−0.1003	0.2866	−0.4002	0.4269	−0.3818	0.2958
3	−0.2970	0.1363	0.1124	−0.3009	0.3276	−0.1847	−0.0512	0.2727	−0.3998
4	−0.2475	0.0114	0.2409	−0.3009	0.1228	0.1539	−0.3245	0.2727	−0.0400
5	−0.1980	−0.0909	0.2891	−0.1852	−0.1134	0.3031	−0.1957	−0.1049	0.3198
6	−0.1485	−0.1704	0.2730	−0.0231	−0.2614	0.2202	0.0985	−0.3063	0.1519
7	−0.0990	−0.2272	0.2088	0.1312	−0.2772	0.0047	0.2824	−0.1552	−0.2079
8	−0.0495	−0.2612	0.1124	0.2392	−0.1732	−0.2013	0.2299	0.1468	−0.2799
9	0	−0.2726	0	0.2778	0	−0.2842	0	0.2937	0
10	0.0495	−0.2612	−0.1124	0.2392	0.1732	−0.2013	−0.2299	0.1468	0.2799
11	0.0990	−0.2272	−0.2088	0.1312	0.2772	0.0047	−0.2824	−0.1552	0.2079
12	0.1485	−0.1704	−0.2730	−0.0231	0.2614	0.2202	−0.0985	−0.3063	−0.1519
13	0.1980	−0.0909	−0.2891	−0.1852	0.1134	0.3031	0.1957	−0.1049	−0.3198
14	0.2475	0.0114	−0.2409	−0.3009	−0.1228	0.1539	0.3245	0.2727	0.0400
15	0.2970	0.1363	−0.1124	−0.3009	−0.3276	−0.1847	0.0512	0.2727	0.3998
16	0.3466	0.2839	0.1124	−0.1003	−0.2866	−0.4002	−0.4269	−0.3818	−0.2958
17	0.3960	0.4543	0.4497	0.4012	0.3276	0.2463	0.1708	0.1091	0.0640

$$n = 18$$

1	2	3	4	5	6	7	8	9	10
1	−0.3862	0.4459	−0.4459	0.4033	−0.3352	0.2577	−0.1836	0.1212	−0.0740
2	−0.3407	0.2885	−0.1311	−0.0712	0.2564	−0.3790	0.4212	−0.3922	0.3177
3	−0.2953	0.1508	0.0852	−0.2788	0.3303	−0.2198	0.0054	0.2175	−0.3678
4	−0.2499	0.0328	0.2164	−0.3025	0.1627	0.0985	−0.2970	0.3031	−0.1197
5	−0.2044	−0.0656	0.2754	−0.2135	−0.0592	0.2804	−0.2430	−0.0178	0.2720
6	−0.1590	−0.1443	0.2754	−0.0712	−0.2230	0.2559	0.0129	−0.2745	0.2329
7	−0.1136	−0.2033	0.2295	0.0771	−0.2780	0.0845	0.2339	−0.2318	−0.0936
8	−0.0681	−0.2426	0.1508	0.1957	−0.2211	−0.1218	0.2704	0.0250	−0.2894
9	−0.0227	−0.2623	0.0524	0.2610	−0.0834	−0.2565	0.1163	0.2496	−0.1523
10	0.0227	−0.2623	−0.0524	0.2610	0.0834	−0.2565	−0.1163	0.2496	0.1523
11	0.0681	−0.2426	−0.1508	0.1957	0.2211	−0.1218	−0.2704	0.0250	0.2894
12	0.1136	−0.2033	−0.2295	0.0771	0.2780	0.0845	−0.2339	−0.2318	0.0936
13	0.1590	−0.1443	−0.2754	−0.0712	0.2230	0.2559	−0.0129	−0.2745	−0.2329
14	0.2044	−0.0656	−0.2754	−0.2135	0.0592	0.2804	0.2430	−0.0178	−0.2720
15	0.2499	0.0328	−0.2164	−0.3025	−0.1627	0.0985	0.2970	0.3031	0.1197
16	0.2953	0.1508	−0.0852	−0.2788	−0.3303	−0.2198	−0.0054	0.2175	0.3678
17	0.3407	0.2885	0.1311	−0.0712	−0.2564	−0.3790	−0.4214	−0.3922	−0.3177
18	0.3862	0.4459	0.4459	0.4033	0.3352	0.2577	0.1836	0.1212	0.0740

$n=19$

Order of polynomial coefficient

1	2	3	4	5	6	7	8	9	10
1	−0.3770	0.4379	−0.4418	0.4046	−0.3416	0.2678	−0.1954	0.1328	−0.0839
2	−0.3351	0.2919	−0.1473	−0.0450	0.2277	−0.3571	0.4126	−0.3984	0.3356
3	−0.2932	0.1631	0.0606	−0.2565	0.3282	−0.2468	0.0549	0.1640	−0.3307
4	−0.2513	0.0515	0.1928	−0.2995	0.1942	0.0472	−0.2625	0.3164	−0.1826
5	−0.2094	−0.0429	0.2599	−0.2340	−0.0100	0.2494	−0.2714	0.0586	0.2098
6	−0.1675	−0.1202	0.2729	−0.1111	−0.1808	0.2730	−0.0620	−0.2226	0.2764
7	−0.1256	−0.1803	0.2426	0.0278	−0.2646	0.1472	0.1705	−0.2695	0.0173
8	−0.0838	−0.2232	0.1798	0.1501	−0.2478	−0.0432	0.2727	−0.0820	−0.2418
9	−0.0419	−0.2490	0.0953	0.2327	−0.1474	−0.2044	0.1967	0.1640	−0.2418
10	0	−0.2576	0	0.2618	0	−0.2666	0	0.2734	0
11	0.0419	−0.2490	−0.0953	0.2327	0.1474	−0.2044	−0.1967	0.1640	0.2418
12	0.0838	−0.2232	−0.1798	0.1501	0.2478	−0.0432	−0.2727	−0.0820	0.2418
13	0.1256	−0.1803	−0.2426	0.0278	0.2646	0.1472	−0.1705	−0.2695	−0.0173
14	0.1675	−0.1202	−0.2729	−0.1111	0.1808	0.2730	0.0620	−0.2226	−0.2764
15	0.2094	−0.0429	−0.2599	−0.2340	0.0100	0.2494	0.2714	0.0586	−0.2098
16	0.2513	0.0515	−0.1928	−0.2995	−0.1942	0.0472	0.2625	0.3164	0.1826
17	0.2932	0.1631	−0.0606	−0.2565	−0.3282	−0.2468	−0.0549	0.1640	0.3307
18	0.3351	0.2919	0.1473	−0.0450	−0.2277	−0.3571	−0.4126	−0.3984	−0.3356
19	0.3770	0.4379	0.4418	0.4046	0.3416	0.2678	0.1954	0.1328	0.0839

$n = 20$

1	2	3	4	5	6	7	8	9	10
1	−0.3684	0.4302	−0.4376	0.4051	−0.3469	0.2768	−0.2063	0.1438	−0.0936
2	−0.3296	0.2943	−0.1612	−0.0213	0.2009	−0.3350	0.4017	−0.4010	0.3498
3	−0.2908	0.1736	0.0384	−0.2346	0.3226	−0.2670	0.0977	0.1135	−0.2907
4	−0.2520	0.0679	0.1702	−0.2931	0.2188	0.0008	−0.2242	0.3165	−0.2298
5	−0.2133	−0.0226	0.2434	−0.2481	0.0335	0.2138	−0.2845	0.1222	0.1419
6	−0.1745	−0.0981	0.2669	−0.1436	−0.1380	0.2758	−0.1236	−0.1609	0.2888
7	−0.1357	−0.1585	0.2497	−0.0161	−0.2419	0.1932	0.1025	−0.2758	0.1106
8	−0.0969	−0.2038	0.2010	0.1052	−0.2580	0.0278	0.2481	−0.1636	−0.1633
9	−0.0582	−0.2340	0.1296	0.1982	−0.1926	−0.1411	0.2414	0.0654	−0.2698
10	−0.0194	−0.2490	0.0447	0.2484	−0.0709	−0.2451	0.0984	0.2399	−0.1278
11	0.0194	−0.2490	−0.0447	0.2484	0.0709	−0.2451	−0.0984	0.2399	0.1278
12	0.0582	−0.2340	−0.1296	0.1982	0.1926	−0.1411	−0.2414	0.0654	0.2698
13	0.0969	−0.2038	−0.2010	0.1052	0.2580	0.0278	−0.2481	−0.1636	0.1633
14	0.1357	−0.1585	−0.2497	−0.0161	0.2419	0.1932	−0.1025	−0.2757	−0.1106
15	0.1745	−0.0981	−0.2669	−0.1436	0.1380	0.2758	0.1236	−0.1609	−0.2888
16	0.2133	−0.0226	−0.2434	−0.2481	−0.0335	0.2138	0.2845	0.1222	−0.1419
17	0.2520	0.0679	−0.1702	−0.2931	−0.2188	0.0008	0.2242	0.3165	0.2298
18	0.2908	0.1736	−0.0384	−0.2346	−0.3226	−0.2670	−0.0977	0.1135	0.2907
19	0.3296	0.2943	0.1612	−0.0213	−0.2009	−0.3350	−0.4017	−0.4010	−0.3498
20	0.3684	0.4302	0.4376	0.4051	0.3469	0.2768	0.2063	0.1438	0.0936

$$n=21$$

Order of polynomial coefficient

1	2	3	4	5	6	7	8	9	10
1	−0.3604	0.4228	−0.4333	0.4051	−0.3514	0.2847	−0.2162	0.1541	−0.1031
2	−0.3243	0.2960	−0.1732	0	0.1757	−0.3132	0.3892	−0.4008	0.3607
3	−0.2883	0.1825	0.0182	−0.2132	0.3144	−0.2817	0.1343	0.0665	−0.2495
4	−0.2523	0.0823	0.1490	−0.2843	0.2373	−0.0404	−0.1844	0.3066	−0.2631
5	−0.2162	−0.0044	0.2265	−0.2571	0.0714	0.1761	−0.2856	0.1726	0.0743
6	−0.1802	−0.0779	0.2584	−0.1698	−0.0964	0.2677	−0.1717	−0.0966	0.2775
7	−0.1441	−0.1380	0.2524	−0.0544	−0.2134	0.2242	0.0364	−0.2587	0.1806
8	−0.1081	−0.1847	0.2159	0.0627	−0.2555	0.0882	0.2067	−0.2175	−0.0743
9	−0.0721	−0.2181	0.1566	0.1610	−0.2216	−0.0756	0.2559	−0.0280	−0.2502
10	−0.0360	−0.2381	0.0821	0.2258	−0.1273	−0.2040	0.1706	0.1730	−0.2111
11	0	−0.2448	0	0.2484	0	−0.2521	0	0.2572	0
12	0.0360	−0.2381	−0.0821	0.2258	0.1273	−0.2040	−0.1706	0.1730	0.2111
13	0.0721	−0.2181	−0.1566	0.1610	0.2216	−0.0756	−0.2559	−0.0280	0.2502
14	0.1081	−0.1847	−0.2159	0.0627	0.2555	0.0882	−0.2067	−0.2175	0.0743
15	0.1441	−0.1380	−0.2524	−0.0544	0.2134	0.2242	−0.0364	−0.2587	−0.1806
16	0.1802	−0.0779	−0.2584	−0.1698	0.0964	0.2677	0.1717	−0.0966	−0.2775
17	0.2162	−0.0044	−0.2265	−0.2571	−0.0714	0.1761	0.2856	0.1726	−0.0743
18	0.2523	0.0823	−0.1490	−0.2843	−0.2373	−0.0404	0.1844	0.3066	0.2631
19	0.2883	0.1825	−0.0182	−0.2132	−0.3144	−0.2817	−0.1343	0.0665	0.2495
20	0.3243	0.2960	0.1733	0	−0.1757	−0.3132	−0.3892	−0.4008	−0.3607
21	0.3604	0.4228	0.4333	0.4051	0.3514	0.2847	0.2162	0.1541	0.1031

$n=22$

1	2	3	4	5	6	7	8	9	10
1	-0.3528	0.4158	-0.4289	0.4047	-0.3550	0.2917	-0.2254	0.1639	-0.1122
2	-0.3192	0.2970	-0.1838	0.0193	0.1521	-0.2917	0.3756	-0.3981	0.3687
3	-0.2856	0.1901	0	-0.1927	0.3043	-0.2917	0.1653	0.0234	-0.2084
4	-0.2520	0.0950	0.1290	-0.2738	0.2509	-0.0768	-0.1447	0.2896	-0.2843
5	-0.2184	0.0119	0.2096	-0.2620	0.1041	0.1382	-0.2776	0.2107	0.0110
6	-0.1848	-0.0594	0.2483	-0.1903	-0.0570	0.2520	-0.2072	-0.0346	0.2495
7	-0.1512	-0.1188	0.2516	-0.0872	-0.1818	0.2425	-0.0241	-0.2258	0.2268
8	-0.1176	-0.1663	0.2258	0.0237	-0.2440	0.1368	0.1562	-0.2459	0.0116
9	-0.0840	-0.2020	0.1774	0.1234	-0.2369	-0.0135	0.2472	-0.1066	-0.1994
10	-0.0504	-0.2257	0.1129	0.1978	-0.1694	-0.1526	0.2159	0.0924	-0.2480
11	-0.0168	-0.2376	0.0387	0.2373	-0.0612	-0.2348	0.0847	0.2309	-0.1094
12	0.0168	-0.2376	-0.0387	0.2373	0.0612	-0.2348	-0.0847	0.2309	0.1094
13	0.0504	-0.2257	-0.1129	0.1978	0.1694	-0.1526	-0.2159	0.0924	0.2480
14	0.0840	-0.2020	-0.1774	0.1234	0.2369	-0.0135	-0.2472	-0.1066	0.1994
15	0.1176	-0.1663	-0.2258	0.0237	0.2440	0.1368	-0.1562	-0.2459	-0.0116
16	0.1512	-0.1188	-0.2516	-0.0872	0.1818	0.2425	0.0241	-0.2258	-0.2268
17	0.1848	-0.0594	-0.2483	-0.1903	0.0570	0.2520	0.2072	-0.0346	-0.2495
18	0.2184	0.0119	-0.2096	-0.2620	-0.1041	0.1382	0.2776	0.2107	-0.0110
19	0.2520	0.0950	-0.1290	-0.2738	-0.2509	-0.0768	0.1447	0.2896	0.2843
20	0.2856	0.1901	0	-0.1927	-0.3043	-0.2917	-0.1653	0.0234	0.2084
21	0.3192	0.2970	0.1838	0.0193	-0.1521	-0.2917	-0.3756	-0.3981	-0.3687
22	0.3528	0.4158	0.4289	0.4047	0.3550	0.2917	0.2254	0.1639	0.1122

$$n = 23$$

Order of polynomial coefficient

1	2	3	4	5	6	7	8	9	10
1	-0.3458	0.4091	-0.4246	0.4038	-0.3580	0.2980	-0.2337	0.1731	-0.1210
2	-0.3143	0.2976	-0.1930	0.0367	0.1302	-0.2709	0.3612	-0.3934	0.3741
3	-0.2829	0.1966	-0.0165	-0.1731	0.2929	-0.2980	0.1912	-0.0157	-0.1682
4	-0.2515	0.1063	0.1103	-0.2622	0.2603	-0.1083	-0.1062	0.2675	-0.2955
5	-0.2200	0.0266	0.1930	-0.2636	0.1319	0.1012	-0.2628	0.2377	-0.0461
6	-0.1886	-0.0425	0.2371	-0.2062	-0.0206	0.2310	-0.2315	0.0224	0.2106
7	-0.1572	-0.1010	0.2481	-0.1151	-0.1490	0.2503	-0.0768	-0.1834	0.2513
8	-0.1257	-0.1488	0.2316	-0.0116	-0.2261	0.1741	0.1025	-0.2529	0.0864
9	-0.0943	-0.1860	0.1930	0.0870	-0.2415	0.0420	0.2220	-0.1660	-0.1323
10	-0.0629	-0.2125	0.1378	0.1670	-0.1987	-0.0978	0.2372	0.0116	-0.2461
11	-0.0314	-0.2285	0.0717	0.2189	-0.1113	-0.2017	0.1497	0.1773	-0.1860
12	0	-0.2338	0	0.2368	0	-0.2398	0	0.2438	0
13	0.0314	-0.2285	-0.0717	0.2189	0.1113	-0.2017	-0.1497	0.1773	0.1860
14	0.0629	-0.2125	-0.1378	0.1670	0.1987	-0.0978	-0.2372	0.0116	0.2461
15	0.0943	-0.1860	-0.1930	0.0870	0.2415	0.0420	-0.2220	-0.1660	0.1323
16	0.1257	-0.1488	-0.2316	-0.0116	0.2261	0.1741	-0.1025	-0.2529	-0.0864
17	0.1571	-0.1010	-0.2481	-0.1151	0.1490	0.2503	0.0768	-0.1834	-0.2513
18	0.1886	-0.0425	-0.2371	-0.2062	0.0206	0.2310	0.2315	0.0224	-0.2106
19	0.2200	0.0266	-0.1930	-0.2636	-0.1319	0.1012	0.2628	0.2377	0.0461
20	0.2515	0.1063	-0.1103	-0.2622	-0.2603	-0.1083	0.1062	0.2675	0.2955
21	0.2829	0.1966	0.0165	-0.1731	-0.2929	-0.2980	-0.1912	-0.0157	0.1682
22	0.3143	0.2976	0.1930	0.0367	-0.1302	-0.2709	-0.3612	-0.3934	-0.3741
23	0.3458	0.4091	0.4246	0.4038	0.3580	0.2980	0.2337	0.1731	0.1210

$n = 24$

1	2	3	4	5	6	7	8	9	10
1	−0.3391	0.4027	−0.4202	0.4027	−0.3604	0.3035	−0.2414	0.1817	−0.1295
2	−0.3096	0.2976	−0.2010	0.0525	0.1097	−0.2507	0.3463	−0.3871	0.3773
3	−0.2801	0.2022	−0.0316	−0.1544	0.2806	−0.3011	0.2128	−0.0510	−0.1295
4	−0.2506	0.1162	0.0928	−0.2499	0.2664	−0.1355	−0.0696	0.2420	−0.2984
5	−0.2212	0.0398	0.1768	−0.2626	0.1552	0.0660	−0.2433	0.2550	−0.0957
6	−0.1927	−0.0270	0.2252	−0.2181	0.0127	0.2065	−0.2461	0.0726	0.1657
7	−0.1622	−0.0844	0.2427	−0.1385	−0.1163	0.2498	−0.1210	−0.1362	0.2576
8	−0.1327	−0.1321	0.2342	−0.0430	−0.2040	0.2009	0.0495	−0.2433	0.1461
9	−0.1032	−0.1703	0.2043	0.0525	−0.2377	0.0896	0.1864	−0.2060	−0.0602
10	−0.0737	−0.1990	0.1578	0.1353	−0.2169	−0.0439	0.2388	−0.0608	−0.2157
11	−0.0442	−0.2181	0.0994	0.1958	−0.1503	−0.1594	0.1938	0.1105	−0.2270
12	−0.0147	−0.2276	0.0339	0.2276	−0.0536	−0.2257	0.0739	0.2226	−0.0952
13	0.0147	−0.2276	−0.0339	0.2276	0.0536	−0.2257	−0.0739	0.2226	0.0952
14	0.0442	−0.2181	−0.0994	0.1958	0.1503	−0.1594	−0.1938	0.1105	0.2270
15	0.0737	−0.1990	−0.1578	0.1353	0.2169	−0.0439	−0.2388	−0.0608	0.2157
16	0.1032	−0.1703	−0.2043	0.0525	0.2377	0.0896	−0.1864	−0.2060	0.0602
17	0.1327	−0.1321	−0.2342	−0.0430	0.2040	0.2009	−0.0495	−0.2433	−0.1461
18	0.1622	−0.0844	−0.2427	−0.1385	0.1163	0.2498	0.1210	−0.1362	−0.2576
19	0.1917	−0.0270	−0.2252	−0.2181	−0.0127	0.2065	0.2461	0.0726	−0.1657
20	0.2212	0.0398	−0.1768	−0.2626	−0.1552	0.0660	0.2433	0.2550	0.0957
21	0.2506	0.1162	−0.0928	−0.2499	−0.2664	−0.1355	0.0696	0.2420	0.2984
22	0.2801	0.2022	0.0316	−0.1544	−0.2806	−0.3011	−0.2128	−0.0510	0.1295
23	0.3096	0.2976	0.2010	0.0525	−0.1097	−0.2507	−0.3463	−0.3871	−0.3773
24	0.3391	0.4027	0.4202	0.4027	0.3604	0.3035	0.2414	0.1817	0.1295

Bibliographical note

The basic approach of studying the behaviour of treatment contrasts over time in this book was developed to cope with the data of Lodge and Roberts given in Chapter 5. The long sequence of 37 observations did not originally appear suitable for analysis by the standard method given by Wishart (1938) and by Rowell and Walters (1976) because of its length. Wishart considered polynomial contrasts that describe the growth curve and analysed these to establish treatment effects. However, the basic method was applicable and fortunately extends to a large number of different types of experiments. For example, there does not appear to be a simple alternative way of examining the relation between random variables as was done in Chapter 9. Also, the ability to include environmental variables is appealing.

The reader who would like to know more about regression with time-series errors should consult Fuller (1976), particularly Chapter 8.

Methods of analysis that take account of correlations between plots were presented by Patterson and Lowe (1970), Battese and Fuller (1972) and Björnsson (1978). The importance of plot correlations has been stressed earlier in the work of Yates (1949, 1954) and Cochran (1939) on rotation and similar experiments.

Formal statistical investigations of growth curves and related topics seems to have commenced with the Potthoff and Roy (1964) and Rao (1965). This was extended and applied by Cole and Grizzle (1966) and Grizzle and Allen (1969). There has been much important work done in this field recently and work is continuing; see, for example, Verbyla (1986), Jennich and Schluchter (1986), Diggle (1988) and Verbyla and Cullis (1990). These methods are complex and do not usually lead to different conclusions from those used here for balanced designs, at least. However, they are important because they provide methods of analysis of data from non-orthogonal designs. Unbalanced designs and poor designs with covariates (Chapter 6) may require these methods. A specialist should then be consulted.

Bibliography

Batschelet, E. (1981) *Circular Statistics in Biology*, Academic Press, London.

Battese, G.E. and Fuller, W.A. (1972) Determination of economic optima from crop rotation experiments. *Biometrics*, **28**, 781–92.

Beattie, B.B., Campbell, J.E. and Roberts, E.A. (1979) Economic and biological factors determining productivity in a young hedgerow planting of apples. *Aust. J. Expt. Agric. Anim. Husb.*, **19**, 753–8.

Björnsson, H. (1978) Analysis of a series of long term grassland experiments with autocorrelated errors. *Biometrics*, **34**, 645–51.

Bliss, C.I. (1967) *Statistics in Biology*, Volume 1, McGraw-Hill, New York.

Bliss, C.I. (1970) *Statistics in Biology*, Volume 2, McGraw-Hill, New York.

Brownlee, H., Scott, B.J., Kearins, R.D. and Bradley, J. (1975) Effects of topdressed superphosphate on the sheep and pasture production of dryland lucerne in central western New South Wales. *Aust. J. Exp. Agric. Anim. Husb.*, **15**, 475–83.

Cochran, W.G. (1939) Long-term agricultural experiments. *J. Roy. Statist. Soc.* (Suppl.), **6**, 104–48.

Cochran, W.G. and Cox, G.M. (1957) *Experimental Designs*, second edition, John Wiley, New York.

Cole, J.W.L. and Grizzle, J.E. (1966) Application of multivariate analysis of variance to repeated measurements experiments. *Biometrics*, **22**, 810–28.

Cox, D.R. (1957) The use of a concomitant variable in selecting an experimental design. *Biometrika*, **44**, 150–8.

Cox, D.R. (1982) Randomisation and concomitant variables in the design of experiments, in *Statistics and Probability: Essays in Honour of C.R. Rao* (eds G. Kallianpur, P.R. Krish Maiah and J.K. Ghosh), North-Holland, Amsterdam, pp. 197–202.

Cox, D.R. and Hinkley, D.V. (1974) *Theoretical Statistics*, Chapman & Hall, London.

Diggle, P.J. (1988) An approach to the analysis of repeated measurements. *Biometrics*, **44**, 959–71.

Draper, N. and Smith, H. (1981) *Applied Regression Analysis*, second edition, John Wiley, New York.

Evans, J.C. and Roberts, E.A. (1979) Analysis of sequential observations with applications to experiments on grazing animals and perennial plants. *Biometrics*, **35**, 687–93.

FitzGerald, R.D. (1979) A comparison of four pasture types for the wheat belt of southern New South Wales. *Aust. J. Exp. Agric. Anim. Husb.*, **19**, 216–24.

Fuller, W.A. (1976) *Introduction to Statistical Time Series*, John Wiley, New York.

Goldstein, H. (1979) *The Design and Analysis of Longitudinal Studies: Their Role in the Measurement of Change*, Academic Press, London.

Green, J.R. and Margerison, D. (1977) *Statistical Treatment of Experimental Data*, Elsevier, Amsterdam.

Grizzle, J.E. and Allen, D.M. (1969) Analysis of growth and dose response curves. *Biometrics*, **25**, 357–81.

Hunter, E.H., Glasbey, C.A. and Naylor, R.G.L. (1984) The analysis of data from germination tests. *J. Agric. Sci.* (Camb), **102**, 207–13.

Jennrich, R.I. and Schluchter, M.D. (1986) Unbalanced repeated-measures models with structured covariance matrices. *Biometrics*, **42**, 805–20.

Johnson, N.L. and Kotz, S. (1970) *Distributions in Statistics: Continuous Univariate Distributions*, Houghton Mifflin, New York.

Lodge, G.M. and Roberts, E.A. (1979) The effects of phosphorus, sulphur and stocking rate on the yield, chemical and botanical composition of natural pasture, north-west slopes, New South Wales, *Aust. J. Exp. Agric. Anim. Husb.*, **19**, 698–705.

McCullagh, P. and Nelder, J.A. (1989) *Generalized Linear Models*, second edition, Chapman & Hall, London.

McLachlan, G.J. and Jones, P.N. (1988) Fitting mixture models to grouped and truncated data via the E.M. algorithm. *Biometrics*, **44**, 571–8.

Patterson, H.D. and Lowe, B.I. (1970) The errors of long-term experiments. *J. Agric. Sci.*, **74**, 53–60.

Potthoff, R.F. and Roy, S.N. (1964) A generalized multivariate analysis of variance model useful especially for growth curve problems. *Biometrika*, **51**, 313–26.

Preece, D.A. (1980) Covariance analysis, factorial experiments and marginality. *The Statistician*, **29**, 97–122.

Rao, C.R. (1965) The theory of least squares when parameters are stochastic and its application to the analysis of growth curves. *Biometrika*, **52**, 447–58.

Roberts, E.A. and Raison, J.M. (1986) Numerical comparison of some test procedures for determining model adequacy when fitting polynomial equations to sequences of observations. *J. Statist. Comput. Simul.*, **24**, 309–17.

Rowell, J.C. and Walters, D.E. (1976) Analysing data with repeated observations on each experimental unit. *J. Agric. Sci.* (Camb), **87**, 423–32.

Sampford, M.R. (1952a) The estimation of response-time distributions: I. Fundamental concepts and general methods. *Biometrics*, **8**, 13–32.

Sampford, M.R. (1952b) The estimation of response-time distributions: II. Multi-stimulus distributions. *Biometrics*, **8**, 307–69.

Sampford, M.R. (1954) The estimation of response-time distributions: III. Truncation and survival. *Biometrics*, **10**, 531–61.

Snedecor, G.W. and Cochran, W.G. (1980) *Statistical Methods*, seventh edition, Iowa State University Press, Ames.

Steel, R.G.D. and Torrie, J.H. (1980) *Principles and Procedures of Statistics*, second edition, McGraw-Hill, New York.

Stuart, A. and Ord, J.K. (1987) *Kendall's Advanced Theory of Statistics*, Vol. 1, fifth edition, Charles Griffin, London.

Verbyla, A.P. (1986) Conditioning the growth curve model. *Biometrika*, **73**, 475–83.

Verbyla, A.P. and Cullis, B.R. (1990) Modelling in repeated measurements experiments. *Applied Statistics*, **39**, 341–56.

Verbyla, A.P. and Cullis, B.R. (submitted for publication). Repeated measures, covariance structures and incomplete data.

Wallace, A.L.C. (1979) Variations in plasma thyroxine concentrations throughout one year in penned sheep on a uniform feed intake. *Aust. J. Biol. Sci.*, **32**, 371–4.

Williams, E.J. (1959) *Regression Analysis*, John Wiley, New York.

Wishart, J. (1938) Growth rate determination in nutrition studies with the bacon pig, and their analysis. *Biometrika*, **30**, 16–28.

Yates, F. (1949) The design of rotation experiments. *Commonw. Bur. Soils Tech. Commun.*, **46**, 142–55.

Yates, F. (1954) The analysis of experiments containing different crop rotations. *Biometrics*, **10**, 324–46.

Yates, J.J., Edye, L.A., Davies, J.G. and Haydock, K.P. (1964) Animal production from a Sorghum almum pasture in south-east Queensland. *Aust. J. Exp. Agric. Anim. Husb.*, **4**, 326–35.

Index